Advance Pra
to the Construction Trades

Finally, a book that tells the truth about the construction trades! *Millennials' Guide to the Construction Trades* encourages young people to find fulfilling careers in the trades. As the co-owner of Heels & Hardhats — and strong supporters of women, LGBT individuals, and diversity in the trades — we can confirm that the construction trades are a great place where anyone willing to work hard can make a good living. For Millennials and others considering the trades, this book will help you choose a trade that is interesting for you, tell you how to get trained (often paid training), and be successful in your construction career. Dream big, work hard, and the world is yours!

—Jackie Richter
President, Heels and Hardhats Contracting
& Endurance Utility Co.
Chicago, Illinois

As President & CEO of an Electrical Supply Company with years of experience supporting the construction industry, *Millennials' Guide to the Construction Trades* is an excellent resource for anyone wanting to join a trade and pursue a solid career in Construction. I have worked and managed different areas of construction and have personally seen so many people who started as apprentices in their respective trades, that have become executives and business owners through the years. This book can guide Millennials in exploring their options and reaching their goals. Congratulations Karl and Jennifer for putting this together to aid the Construction Champions of our future.

—Sandra Escalante
President & CEO, Laner Electric Supply Co., Inc.
Richmond, California

The alarm clock sounds and you turn on the lights. Hot water from the shower wakes you up as coffee brews in the kitchen. As you start the day, these and hundreds of other small miracles of the modern age are possible due to the hard work of skilled construction tradesmen and women. Electricians, plumbers, roofers, HVAC technicians, and a myriad of other skilled workers come together every day to build and maintain the modern world around us. The *Millennials' Guide to the Construction Trades* is a great place to start for people considering entering the trades. It uses simple, straight-forward language to cover all the topics that anyone interested in the trades should be curious about and provides good, solid, life guidance that should allow them to be successful in any trade career. I am a strong advocate for young people entering the trades and encourage folks that do not see college as a career path to consider the trades as a great way to provide for their families and find stable, rewarding work for a lifetime.

—**Mike Mangin**
Facility Manager, Caterpillar, Inc.
Atlanta, Georgia

The construction trade can greatly benefit from what the Millennials bring to the table: new and innovative ideas mixed in with the knowledge and skills of what the veterans can teach them. As a supervisor in structural concrete, I want to give this book to every new worker on our team! *Millennials' Guide to the Construction Trades* is a great overview of the many opportunities and benefits for a career in the trades. It's worth it for the success tips alone! I wish I had this book when I was first starting out.

—**Rick Bird**
Supervisor, Structural Concrete
Portland, Oregon

As an architect, I work with many aspects of the construction trades. As a parent, I appreciate this amazing opportunity for young people to learn about the construction trades and create a wonderful career for themselves. Millennials, check out this book–it is packed with so much information about opportunities in the construction industry and can literally change your life!

—Heidi M Bolyard
Architect and Owner, Simplified Living Architecture
Dublin, Ohio

Construction isn't your typical "Sign up for the class/Read a textbook/ Study and ace the test" career. It involves a lot of do-it-yourself learning, always asking questions, and always being open to hard work. The *Millennials' Guide to the Construction Trades* gives you the best introduction to the world of building. It discusses who does construction work, and what it takes to be successful in the field.

—Remi Farnan
Construction Project Manager, Sciame Construction
New York, New York

Wow! What a resource! As a property manager for private families I've worked with more vendors and tradesmen and women than I can count. I wish each of them had this book. My recommendation: Focus on Sections III (Success Tips for Construction Workers) and IV (Common Challenges and Opportunities in the Trades). Read every word of these pages; they will serve you well. I look forward to working with anyone using this guide and wish each of you a lifetime of success. Success will be yours with this book in your tool belt.

—Jennifer Trepeck
Custom Property Management LLC
New York, New York

Millennials' Guide to the Construction Trades is a fantastic resource for anyone considering the trades. I am in construction sites all day, every day, and I'm familiar with the benefits and challenges of a career in construction. This book will help you find your place in the exciting and adventurous world of construction.

—**Tracy Passarella**
Construction Project Manager
New York, New York

Thanks to Jennifer Wisdom and Karl Hughes for giving a voice to the voiceless. As the founder of the National Association of Black Women in Construction, and the voice for Millennial Black women in Construction, this book, *Millennials Guide to the Construction Trades*, is phenomenal. Everybody is not going to go to college and for those individuals, old and young, who choose a different path, in my experience, there is no greater path than the path of Construction. Personally, I hold two General Contracting Licenses, an undergraduate degree in Accounting, and a Master's in finance. And when I started in this industry, I did not listen to the naysayers who laughed at me and told me I was crazy to go into construction. Now, 40+ years later, they are no longer laughing. As the first Black woman contractor in Florida operating in the level that we do, when I received your book and began to read I automatically knew it was exactly what the industry needs to begin to change the dialogue and the perception and image and the brand for the Construction Industry. We are excited about the book, we will be using it throughout our organization, and we want to encourage and challenge the reader to live in the pages of this book.

—**Ann McNeill**
President/CEO of MCO Construction, the first Black
Woman-owned construction company in Florida and Founder,
National Association of Black Women in Construction

MILLENNIALS'
GUIDE TO THE
CONSTRUCTION
TRADES

MILLENNIALS'
GUIDE TO THE
CONSTRUCTION
TRADES

What No One Ever Told You
About a Career in Construction

KARL D. HUGHES
JENNIFER P. WISDOM

Published by Winding Pathway Books

WINDING PATHWAY BOOKS

ISBN (print): 978-1-7330977-6-5

ISBN (e-book): 978-1-7330977-7-2

Book design by: Deana Riddle at Bookstarter
and Jerry Dorris at AuthorSupport.com

Cover design by Brian Sisco at 115 Studios and Diego G. Diaz

Photo credit: Diego G. Diaz

For more information or bulk orders, visit: www.leadwithwisdom.com

Printed in the United States of America

ACKNOWLEDGMENTS

L ike construction, completing this book was a group effort. We are very appreciative of colleagues who reviewed earlier versions of this book, including Cassandra Blake, Remi Farnan, Alison Feuer, Mike Mangin, and Tracy Passarella. We also appreciate additional support from Thomas Chernick and Diego G. Diaz.

Karl would like to acknowledge his colleagues at the New York City District Council of Carpenters Training Center for their generous insight and input on this project. Special thanks to Pete Bennett, Andrew Irenze, Lou Rioux, and Ron Zimmeman for their support. Thanks to the following Millennials he had the pleasure of mentoring through the years who shared their perspective as they follow their paths in the construction trades: Ed Cerracchio, Joseph Romano, and Steven Bain. And as always, my sincere thanks to my best friend Patricia S. Hughes.

Jennifer would like to share enormous appreciation of her amazing team of supporters and readers: Carolynn Ananian, Katrina Amaro, Tara Amato, Lourdes Blanco, Prea Gulati, Robyn Hatcher, Lisa D. Jenkins, Caroline Mays, Priti Mehta, David Mohammed, William Monks, Jean Pak, Rochelle Rice, Jenna Van Leeuwen, Tildet Varon, and Linda Warnasch. Thanks also especially to supreme administrator Cassandra Blake.

We appreciate cover design by Brian Sisco at 115 Studios and Diego G. Diaz, copyediting by Margaret McConnell, and additional cover and interior design by Deana Riddle at BookStarter.

TABLE OF CONTENTS

FOREWORD

Working in the construction industry (specifically in ceramic tile installation) has taken me from growing up in a tiny farming community in northern Minnesota to living and working in some of the coolest places in the United States and around the world. I've lived and worked in New York City; Monterrey, California; and Minneapolis, Minnesota, just to name a few. I've been able to travel the world, going as far east as the Republic of Georgia, backpacking from El Salvador to Panama City, and traveling across the country from California to Florida with my wife and daughter in our fifth wheel. I'm currently living in Bradenton, Florida, within walking distance from the beach. If I wanted to, I could call no less than a dozen industry friends throughout the nation and secure a position working for them.

Who would have thought that my ticket to all my rich and exciting adventures was directly related to my apprenticeship in the ceramic tile installation industry! But travel is just one of the reasons that I always encourage Millennials to consider working in the construction industry for a rewarding career. There are many other reasons to give serious consideration to the construction industry for a career.

I'm excited that you are reading this book, which will assist you in your decision to pursue a career that can be rewarding in many ways!

Don't believe the negative misconceptions surrounding working with your hands until you do your own research. Too many people will tell you that working in the construction industry is the last resort option. They will tell you that you will not earn a good living. They will tell you that only the uneducated or people with a history of breaking the law are working in the trades. Well, I'm here to tell you not to believe the negative naysayers! If you have an inclination that you may be happy working with your hands, you owe it to yourself to read this book and continue to do research into which trade specifically would be best for you.

The truth is that learning a trade can be very rewarding, financially as well as mentally and emotionally. A few of the benefits are: enjoying the satisfaction that comes from building something from nothing, staying physically active all day at work, getting paid to learn a trade vs. spending money on general education that may or may not lead to a career. These are just a few of the benefits and joys of a career in the construction industry.

There are many different types of careers to consider when thinking about the construction trades. This book does a great job of bringing to light the differences between your options. I encourage you to take your time when attempting to identify which trade speaks to you. Some trades will give you a variety of tasks to accomplish, often changing daily. Other trades will be highly specialized and focus more on daily routine and consistency. If you find yourself confused and undecided, my advice is to choose one of the more general trades, so you can get some well-rounded training while you are getting paid to learn new skills. Two trades that would allow this type of general training and a wide variety of tasks are Carpenter (Trade 4) and Laborer (Trade 14). My chosen trade and personal favorite, Flooring Installer/ Tile and Marble Setter (Trade 9), will also lead to knowledge of other skills like waterproofing and building showers. Often starting with general trades will lead to your focusing on a specific trade that speaks to you and allows you to specialize in a more specific trade.

After you read through what the trades have to offer, Section II of this book focuses on the skills and abilities construction workers need. Do not be discouraged if you feel you do not possess all of these skills and abilities, because you will develop and hone these during and after your apprenticeship. These do reflect what is needed to be a skilled construction worker, as well as what your potential employers are looking for. Section III provides success tips that will serve you well. These sections will be great continued reference material as you go through your apprenticeship and onto the job site. If you desire more responsibilities, these are the things you need to continue to improve during your apprenticeship. Section IV covers some of the challenges — as well as opportunities – you are likely to encounter during your career. You'll want to keep this book handy for a long time.

Construction business owners everywhere are struggling to fill positions with reliable help. In fact, you are interviewing your potential employer just as much as they are interviewing you. Not every company will possess the culture that you desire, but with a little effort and in time I am confident that you can find an employer who will value you and what you have to offer. Very best wishes to a successful career in the construction trades!

Luke Miller
Tile Money Podcast
TileMoney.com

INTRODUCTION

Construction trades have been in existence since humans first sought shelter from the elements, yet today the construction field is a career that is significantly underappreciated. Construction trades have developed from crude work with chisels and hammers to the building of aqueducts and temples, to today's spectacular construction of bridges and skyscrapers. As the world continues to develop, it is clear that the construction trades are here to stay. The U.S. has one of the largest construction markets in the world, spending $977 billion in 2019 with about 11.2 million people employed in construction. What a great opportunity for Millennials! With constant improvements in technology and the insatiable need for new development, structures that were only fantasies even a few decades ago are already being completed around the world on a constant basis.

Do you know what you want to do when you grow up? Kids wonder about this all the time, and sometimes we are not sure what we want to do well into our 20s, 30s, or even 40s! High schools may not be fully informed about careers in blue-collar fields, such as construction. High schools often advise their students to continue their education by going to college. The recommendation is that a young student will need this education in order to prosper in a career and in life. Sometimes counselors suggest that, without a college degree, you will end up in a dead-end job without any prospects for a financially secure life. This is a powerful argument to go to college! Unfortunately, the cost of that education continues to increase so that many times it causes financial insecurity rather than the promised prosperity. Many, many people have skipped college or entered the construction trades after some college to find good wages, fulfilling work, and a satisfying career.

The construction trades rapidly embrace new ideas and improvements, and as such, are a great target for Millennials. Because of the high cost of materials and labor and the pressure of time schedules, there is a constant demand for newer, better, safer, cheaper ways to build — and more Millennials with skills and persistence to build them. Unions and other societal pressures have pushed the trades to welcome women, ethnic minorities, and immigrants to its ranks of proud workers.

Construction is filled with uncertainty, noise, difficulty, and some of the most challenging and rewarding work ever accomplished by humans. The construction trades are very different from the corporate world, and very different from careers with an emphasis on higher education. In many ways, the construction trades provide comfort for those other occupations and corporations to do the work and conduct the businesses that they do.

Construction can be a good fit for Millennials who are comfortable with these aspects of the trade:

- **All construction work is temporary.** While some projects can be enormous in scope, other projects are very brief. The only constant in construction is that all work being done by the construction trades is to provide the skills necessary to build a structure or to build for the benefit of others. Once that project is complete, it is no longer a construction project, and the trades are no longer involved.

- **Construction doesn't require you to be good in academics.** If school wasn't your thing, construction might be a good fit. The construction trades provide a way to earn a living, even with just a basic education. You do have to be able to pick up skills and be open to learning on the job, but it's very different from a classroom environment.

- **Construction doesn't judge who you are, as long as you're good at what you do.** For many Millennials, entering into the construction trades can provide a fresh start from a previous career or life situation.

◆ **For those who like to be outside, for those who do not want to work at a desk, and for those who do not want a mundane repetitive job, construction may be a great choice.** Much of the work that trades perform is done outside, on anything from a small house to a huge skyscraper or a superhighway.

◆ **If you like to build and create things, construction may be for you.** If you enjoy the satisfaction of getting things done right and having a product to look at when you're done, construction could be a good fit. There are lots of ways in construction to create things, from working with wood, concrete, metal, glass, or equipment.

There are many reasons for Millennials to choose a career in construction. Whatever brings you to the construction field, we are hopeful that *Millennials' Guide to the Construction Trades* will help you be successful!

Karl D. Hughes
Jennifer P. Wisdom
July 2020

HOW TO USE THIS BOOK

This is not a book best read cover to cover. We encourage you to review the table of contents and identify something that interests you. Turn to those pages to start reading!

The book starts with an introduction to 25 **trades**, a brief history of each trade, what people in those trades do, the kinds of skills needed for each trade, and information on compensation and employment outlook (note: nearly all construction trades have excellent employment outlook!). We pull heavily from the U.S. Department of Labor's Occupational Outlook Handbook to provide the most recent data available. If you're considering a job in the trades, take a look at the variety of opportunities, and see what looks good to you.

Next, we describe **skills and abilities** common across all the trades, like mechanical skills, ability to overcome fears, and determination. It would be helpful for anyone in the trades to review these and identify which are strengths, and which you might want to work on.

Success tips are an important aspect of any profession, and the construction trades are no exception. These tips give you pointers for how to be successful at work. Reviewing these tips can help you understand your co-workers, your boss, and any challenges you may encounter at work in the construction trades.

Finally, we describe **common challenges and opportunities** that you might encounter in the construction trades. Every challenge is, of course, also an opportunity! From being a woman, an under-represented racial/ethnic minority, or an LGBTQ person in the trades, to how to interview, ask for a raise, or when you think it might be time to move on—this section addresses all of these. Note that each challenge/opportunity gives you a chance to practice important skills, such as knowing what you want, communicating with others, and choosing the life you want. For each challenge and opportunity, we provide at least 5 solutions. You'll see some solutions repeated across different challenges because they're likely to be helpful for many problems. Sometimes, you can see success after trying just one option. For complex challenges, you may want to attempt several interventions at the same time. It's helpful if you approach problems with curiosity; it's not likely personal, so see what you can figure out to solve the problem.

We also include a **For Further Reading** section that includes references for all of the trades as well as additional resources for further reading on the values, qualities, and challenges/opportunities you might encounter.

As in all areas of work, it's important to remember a few basic rules of work that will never steer you wrong:

1. Never say anything bad about anyone at work to anyone at work. (Do your venting at home or with friends.)

2. Be honest *and* diplomatic with everyone, including yourself.

3. Be patient. Sometimes people are working on your behalf to make things better and you don't even know it.

4. Be curious about yourself and seek constant self-improvement.

5. Remember that we all have struggles. Be kind and respectful.

Each of you reading this book is a unique person with talents to share with the world. Our hope is that this book can make it easier for you to do so. Good luck in the construction trades!

SECTION I

Descriptions of the trades

TRADE 1

Boilermaker

What do nuclear power plants, shipyards, and refineries have in common? They all use storage containers made by boilermakers. A boilermaker is a tradesperson who cuts, bends, and assembles steel, iron, or copper into boilers and other large containers to hold hot gas or liquid. Sound cool? Read on!

1. **History of boilermaking:** Boilermaking evolved from industrial blacksmithing. In the 1800s, the advent of steam power created a need for boilers to heat water to create steam. Boilers were made of plates of metal or tubes that were cut, bent and shaped by the boilermakers. Boilers are also used to generate the electricity we all use by turning steam turbines at power plants around the world. The railroad industry relied on boilers to drive the steam locomotives. Boilermakers installed most of the piping used in the hydroelectric dams still in operation today. Boilermakers also became more involved in shipbuilding and engineering when there was significant change from wood to iron, and later steel, as construction material.

2. **What boilermakers do:** Boilermakers assemble, install, maintain, and repair boilers, closed vats, and other large vessels or containers that hold liquids and gases.

3. **Work environment:** Boilermakers do physically demanding and dangerous work. They often work outdoors in all types of weather, including in extreme heat and cold. Because boilers, storage tanks, and pressure vessels are large, boilermakers often

work at great heights. When working on a dam, for example, they may be hundreds of feet above the ground. Boilermakers also work in cramped quarters inside boilers, vats, or tanks that are often dark, damp, and poorly ventilated.

4. **Education needed**: A high school diploma or equivalent is typically required.

5. **Training needed:** Boilermakers train through an apprenticeship program (see #6). Programs typically provide instruction on towers, vessels, and furnaces; safety analysis and pre-job safety checklists; and up-to-date techniques for cutting and fitting gaskets, base metal preparation, and welding basics.

6. **Apprenticeships:** Some unions and contractor associations sponsor apprenticeship programs. Apprenticeship applicants who have previous welding or other related experience—such as working as a pipefitter, millwright, sheet metal worker, or welder—may have priority over applicants without any experience. In addition, those with experience or education may qualify for a shortened apprenticeship. During the apprenticeship, workers learn how to use boilermaker tools and equipment on the job. They also learn about metals and installation techniques, blueprint reading and sketching, safety practices, and other topics. Apprenticeship programs typically last 4 years.

7. **After apprenticeship**: When boilermakers finish an apprenticeship, they are considered to be journey-level workers.

8. **License/certification:** Some states require boilermakers to have a license; licensure requirements typically include work experience and passing an exam. Employers may require or prefer that boilermakers hold certification from the National Center for Construction Education and Research (NCCER). Welding certification may also be helpful.

9. **Competencies needed**: The following are required competencies for boilermakers.

    a. Mechanical skills. Boilermakers use and maintain a variety of equipment, such as hoists and welding machines.

    b. Physical stamina. Boilermakers spend many hours on their feet while lifting heavy boiler components.

    c. Physical strength. Boilermakers must be able to move heavy vat components into place.

    d. Unafraid of confined spaces. Boilermakers often work inside boilers and vats.

    e. Unafraid of heights. Some boilermakers work at great heights. While installing water storage tanks, for example, workers may need to weld tanks several stories above the ground.

10. **Union/non-union:** Boilermakers often belong to The International Brotherhood of Boilermakers, Iron Ship Builders, Blacksmiths, Forgers and Helpers, which is affiliated with the American Federation of Labor and Congress of Industrial Organizations (AFL-CIO) and Central Labor Councils (CLCs). Unions can provide health insurance, job security, pensions, and representation.

11. **Compensation:** In 2019, median pay for boilermakers was $63,100 annually, or $30.34 per hour, not including benefits such as health insurance and retirement. Starting pay is closer to $39,840. Top earners earn $94,440 annually or more. Note that generally, union jobs pay higher than non-union jobs, and cities pay higher than rural areas. In 2018, there were 14,500 boilermaker jobs in the U.S.

12. **Employment outlook:** Employment of boilermakers is projected to grow 6 percent from 2018 to 2028, about as fast as the average for all occupations. The need to install, replace, and maintain boiler parts—such as boiler tubes, heating elements,

and ductwork—is an ongoing process that will require the work of boilermakers.

See also: **Trade 13:** Ironworker
Trade 16: Millwright
Trade 19: Plumber, pipefitter, and steamfitter
Trade 21: Sheet metal worker

TRADE 2

Bricklayer/Mason

Bricklayers and masons are one of the world's oldest construction trades. From setting individual bricks to finishing stone buildings, bricklayers and masons create beautiful structures.

1. **History of masonry:** Brick is humans' oldest manufactured product. Some of the world's most impressive architectural creations, such as the Egyptian pyramids, the Colosseum in Rome, the Taj Mahal in India, and the Great Wall of China, were all built by masons. Architects and builders have chosen masonry for its beauty, versatility, and durability. Masonry structures can withstand the normal wear and tear of centuries.

2. **What masons do:** Masonry workers use bricks, concrete blocks, concrete, and natural and synthetic stones to build structures.

3. **Work environment:** The work is physically demanding because masons lift heavy materials and often must stand, kneel, or bend for long periods of time. Poor weather conditions may reduce work activity because masons usually work outdoors.

4. **Education needed**: A high school diploma or equivalent is typically required for most masons.

5. **Training needed:** Many technical schools offer programs in masonry. Some people take courses before being hired, and some take them later as part of on-the-job training. Some workers start out working as construction laborers and helpers before becoming a mason.

6. **Apprenticeships:** Most masons learn the trade through apprenticeships and on the job, working with experienced masons. Several groups, including unions and contractor associations, sponsor apprenticeship programs. Apprentices learn construction basics, such as blueprint reading; mathematics for measurement; building code requirements; and safety and first-aid practices.

7. **After apprenticeship:** After completing an apprenticeship program, masons are considered journey workers and are able to perform tasks on their own. The Home Builders Institute (HBI) and the International Masonry Institute (IMI) offer pre-apprenticeship training programs for eight construction trades, including masonry.

8. **License/certification:** None needed.

9. **Competencies needed:** The following are required competencies for masons.

   a. **Color vision.** Terrazzo workers need to be able to distinguish between small variations in color when setting terrazzo patterns in order to produce the finished product.

   b. **Dexterity.** Masons repeatedly handle bricks, stones, and other materials and must place bricks and materials with precision.

   c. **Hand–eye coordination.** Masons apply smooth, even layers of mortar; set bricks; and remove any excess before the mortar hardens.

   d. **Physical stamina.** Brickmasons must keep a steady pace while setting bricks. Although no individual brick is extremely heavy, the constant lifting can be tiring.

   e. **Physical strength.** Workers should be strong enough to lift more than 50 pounds. They carry heavy tools, equipment, and other materials, such as bags of mortar and grout.

f. Unafraid of heights. Masons often work on scaffolding, so they should be comfortable working at heights.

10. **Union/non-union** Many bricklayers and masons join the International Union of Bricklayers and Allied Craftworkers (BAC). The national trade association representing masons is the Mason Contractors Association of America (MCAA). Unions can provide health insurance, job security, pensions, and representation.

11. **Compensation**: In 2019, median pay for bricklayers was $46,500 annually, or $22.35 per hour, not including benefits such as health insurance and retirement. Starting annual pay is closer to $30,250. Top earners earn $78,250 annually or more. Note that generally, union jobs pay higher than non-union jobs, and cities pay higher than rural areas. In 2018, there were 298,000 bricklayer jobs in the U.S.

12. **Employment outlook**: Employment of masonry workers is projected to grow 11 percent from 2018 to 2028, much faster than the average for all occupations. Population growth will result in the construction of more schools, hospitals, homes, and other buildings. Workers with experience in construction should have the best job opportunities.

See also: Trade 9: Flooring Installer/Tile and Marble Setter
Trade 18: Plasterer/Stucco Mason

TRADE 3

Cabinetmaker/ Woodworker

A significant subtype of carpentry is cabinetmaking and woodworking. These are individuals who create cabinets, furniture, and other highly detailed structures. Cabinetmakers and woodworkers can work as independent craftspeople or contribute to building within a manufacturing setting.

1. **History of cabinetmaking/woodworking:** Woodworking is the art and trade of cutting, working, and joining timber. Ancient Egyptian drawings and artifacts going back to 2000 B.C. demonstrate wood furnishings such as beds, chairs, tables, beds, and chests.

2. **What cabinetmakers and woodworkers do:** Woodworkers manufacture or create a variety of products such as cabinets and furniture, using wood, veneers, and laminates.

3. **Work environment:** Most woodworkers work in manufacturing plants, as opposed to carpenters, who typically work at construction sites. Although working conditions vary, some woodworkers may encounter machinery noise and wood dust.

4. **Education needed:** A high school diploma or equivalent is typically required.

5. **Training needed:** Education is helpful, but woodworkers are trained primarily on the job, where they learn skills from

experienced workers. Beginning workers are given basic tasks, such as placing a piece of wood through a machine and stacking the finished product at the end of the process. As they gain experience, new woodworkers perform more complex tasks with less supervision. In about 1 month, they learn basic machine operations and job tasks. Becoming a skilled woodworker often takes several months or even years. Skilled workers can read blueprints, set up machines, and plan work sequences. Although some entry-level jobs can be learned in less than 1 year, becoming fully proficient generally takes several years of on-the-job training. The ability to use computer-controlled machinery is becoming increasingly important.

6. **License/certification**: Although not required, becoming certified can demonstrate competence and professionalism. It also may help a candidate advance in the profession. The Woodwork Career Alliance of North America offers a national certificate program, with five progressive credentials, which adds a level of credibility to the work of woodworkers.

7. **Competencies needed**: The following are required competencies for cabinetmakers.

    a. **Detail oriented.** Woodworkers must pay attention to details in order to meet specifications and to keep themselves safe.

    b. **Dexterity.** Woodworkers must make precise cuts with a variety of hand tools and power tools, so they need a steady hand and good hand-eye coordination.

    c. **Math skills.** Knowledge of basic math and computer skills are important, particularly for those who work in manufacturing, in which technology continues to advance. Woodworkers need to understand basic geometry in order to visualize how a three-dimensional wooden object, such as a cabinet or piece of furniture, will fit together.

d. **Mechanical skills.** The use of hand tools, such as screwdrivers and wrenches, is required to set up, adjust, and calibrate machines. Modern technology systems require woodworkers to be able to use computers and other programmable devices.

e. **Physical stamina.** The ability to endure long periods of standing and repetitive movements is crucial for woodworkers, who often stand all day performing many of the same functions.

f. **Physical strength.** Woodworkers must be strong enough to lift bulky and heavy pieces of wood.

g. **Technical skills.** Woodworkers must understand and interpret design drawings and technical manuals for a range of products and machines.

8. **Union/non-union:** Woodworkers and cabinetmakers in the U.S. are typically represented by the United Brotherhood of Carpenters and Joiners of America (UBC) or the Amalgamated Union of Cabinet Makers (AUCM). Unions can provide health insurance, job security, pensions, and representation.

9. **Compensation:** In 2019, median pay for cabinetmakers was $32,690 annually, or $15.72 per hour per hour, not including benefits such as health insurance and retirement. Starting pay is closer to $22,830. Top earners earn $54,140 or more annually. Note that generally, union jobs pay higher than non-union jobs, and cities pay higher than rural areas. In 2018, there were 272,200 cabinetmaker jobs in the U.S.

10. **Employment outlook:** Employment of woodworkers is projected to show little or no change from 2018 to 2028. Those who have advanced skills, including the ability to use computer-controlled machinery, will have the best job opportunities in manufacturing industries.

See also: **Trade 4:** Carpenters
Trade 7: Drafter
Trade 19: Plumber, Pipefitter, and Steamfitter
Trade 22: Surveyor

TRADE 4

Carpenter

Carpenters work with wood. They are experts at using wood for establishing building structures, framing walls, or crown molding to emphasize the corners in a room. Carpenters focus on how to shape and join wood to create useful and beautiful structures.

1. **History of carpentry:** Carpentry is the art and trade of cutting, working, and joining timber. The term includes both structural timberwork in framing and items such as doors, windows, and staircases.

2. **What carpenters do:** Carpenters construct, repair, and install building frameworks and structures made from wood and other materials. Carpenters can complete large projects, such as framing houses or building bridges; moderately large projects such as installing drywall; or detailed finish projects such as kitchen cabinetry. Carpenters may use a variety of hand or power tools to cut and shape wood, plastic, fiberglass, drywall and other substances. Carpenters then fasten these materials with nails, screws, staples, and adhesives. The result is something entirely new that is more useful than a piece of wood.

3. **Work environment:** Carpenters can work indoors or outdoors, depending on the job. Carpenters typically work at construction sites, as opposed to woodworkers, who work in manufacturing plants.

4. **Education needed**: A high school diploma or equivalent

is typically required. High school courses in mathematics, mechanical drawing, and general vocational technical training are considered useful.

5. **Training needed:** Some technical schools offer associates degrees in carpentry. The programs vary in length and teach basics and specialties in carpentry.

6. **Apprenticeships:** Carpenters typically learn on the job and through apprenticeships and learn the proper use of hand and power tools on the job. They often begin doing simpler tasks under the guidance of experienced carpenters. For example, they start with measuring and cutting wood, and learn to do more complex tasks, such as reading blueprints and building wooden structures. Several groups, such as unions and contractor associations, sponsor apprenticeship programs. For each year of a typical program, apprentices must complete 144 hours of technical training and 2,000 hours of paid on-the-job training. Apprentices learn carpentry basics, blueprint reading, mathematics, building code requirements, and safety and first aid practices. They also may receive specialized training in creating and setting concrete forms, rigging, welding, scaffold building, and working within confined workspaces. All carpenters must pass the Occupational Safety and Health Administration (OSHA) 10- and 30-hour safety courses.

7. **After apprenticeship:** After completing an apprenticeship program, carpenters are considered journey workers and are able to perform tasks on their own.

8. **License/certification:** Many carpenters need a driver's license or reliable transportation, since their work is done on jobsites. Carpenters do not need certification for the job. However, there are certificate programs that teach basics for carpenters interested in completing an apprenticeship, such as the Pre-Apprenticeship Certificate Training (PACT) offered by the Home Builders Institute. Other programs offer certifications by specialty. For

example, the National Association of the Remodeling Industry (NARI) offers various levels of certificates for remodeling

9. **Competencies needed**: The following are required competencies for carpenters.

   a. **Business skills**. Self-employed carpenters must bid on new jobs, track inventory, and plan work assignments.

   b. **Detail oriented**. Carpenters make precise cuts, measurements, and modifications. For example, properly installing windows and frames provides greater insulation to buildings.

   c. **Dexterity**. Carpenters use many tools and need hand-eye coordination to avoid injury or damaging materials. For example, incorrectly striking a nail with a hammer may cause damage to the nail, wood, or oneself.

   d. **Math skills**. Carpenters frequently use basic math skills to calculate area, precisely cut material, and determine the amount of material needed to complete the job.

   e. **Physical strength**. Carpenters use heavy tools and materials that can weigh up to 100 pounds. Carpenters also need physical endurance; they frequently stand, climb, or bend for many hours.

   f. **Problem-solving skills**. Carpenters may need to modify building material and make adjustments onsite to complete projects. For example, if a prefabricated window that is oversized arrives at the worksite, carpenters shave the framework to make the window fit.

10. **Union/non-union:** Carpenters in the U.S. typically join the United Brotherhood of Carpenters and Joiners of America (UBC). Unions can provide health insurance, job security, pensions, and representation.

11. **Compensation:** In 2019, median pay for carpenters was $48,330

annually, or $23.24 per hour, not including benefits such as health insurance and retirement. Starting annual pay is closer to $30,170. Top earners earn $84,690 annually or more. Note that generally, union jobs pay higher than non-union jobs, and cities pay higher than rural areas. In 2018, there were 1,006,500 carpenter jobs in the U.S.

12. **Employment outlook**: Carpenters are involved in many phases of construction and may have opportunities to become first-line supervisors, independent contractors, or general construction supervisors. Employment of carpenters is projected to grow 8 percent from 2018 to 2028, faster than the average for all occupations. Increased levels of new homebuilding and remodeling activity will require more carpenters.

See also: **Trade 3:** Cabinetmaker/Woodworker
Trade 7: Drafter
Trade 19: Plumber, Pipefitter, and Steamfitter
Trade 22: Surveyor
Trade 25: Construction Manager/Project Manager

TRADE 5

Construction and Building Inspector

Think you know your stuff? A construction and building inspector has to know everything there is to know about construction, building codes, zoning regulations, and contract specifications.

1. **History of construction and building inspectors:** Building codes began in ancient times, as early codes indicated liability for dam-builders if their dam breaks and harms others' property, and the Bible indicated that roofs must have barriers to keep people from falling off. After the Great Fire of London in 1666, building regulations started to flourish to maintain safety. As building codes and zoning regulations became more complex, state, county, and city governments developed a need for inspectors to review buildings in progress to ensure they were being built consistent with these codes and regulations. In the 1970s, home buyers began hiring general building inspectors to inspect their homes. Within 20 years, home inspections became standard for home buyers.

2. **What construction and building inspectors do:** Construction and building inspectors ensure that construction meets building codes and ordinances, zoning regulations, and contract specifications. Inspectors review plans to ensure they meet building codes, local ordinances, zoning regulations, and contract specifications; approve building plans;

3. monitor construction sites periodically; inspect plumbing, electrical, and other systems to ensure that they meet code; verify alignment, level, and elevation of structures to ensure building meets specifications; and issue violation notices and stop-work orders until building is compliant. Specialized inspectors can focus on buildings (or types of structures, such as steel or reinforced concrete), coating (such as on bridges and pipelines), electrical work, elevators, mechanical, plumbing, and public works (such as water and sewer systems, dams, and highways).

4. **Work environment:** Construction and building inspectors spend considerable time inspecting worksites, alone or as part of a team. Some inspectors may have to climb ladders or crawl in tight spaces. Most work full time during regular business hours.

5. **Education needed**: Most employers require construction and building inspectors to have at least a high school diploma and work experience in construction trades. Inspectors also typically learn on the job. Some employers may seek candidates who have studied engineering or architecture or who have a certificate or an associate's degree that includes courses in building inspection, home inspection, construction technology, and drafting. Many inspectors previously held jobs as carpenters, plumbers, electricians, or surveyors.

6. **Training needed:** In general, construction and building inspectors receive much of their training on the job, although they must learn building codes and standards on their own. Working with an experienced inspector, they learn about inspection techniques; codes, ordinances, and regulations; contract specifications; and recordkeeping and reporting duties. Training also may include supervised onsite inspections. Many community colleges offer programs in building inspection technology. Courses in blueprint reading, vocational subjects, algebra, geometry, and writing are also useful. Courses in business management are helpful for those who plan to run their own inspection business.

7. **Apprenticeships:** Some unions or other programs offer apprenticeships to provide education and on-the-job training for construction inspectors.

8. **License/certification:** Many states and local jurisdictions require some type of license or certification. Some states have their own licensing programs, and others may require certification by associations such as the International Code Council (ICC), the International Association of Plumbing and Mechanical Officials (IAPMO), the International Association of Electrical Inspectors (IAEI), and the National Fire Protection Association (NFPA).

9. **Competencies needed:** The following are required competencies for construction and building inspectors.

    a. **Communication skills.** Inspectors must explain problems they find in order to help people understand what is needed to fix the problems. In addition, they need to provide a written report of their findings.

    b. **Craft experience.** Inspectors perform checks and inspections throughout the construction project. Experience in a related construction occupation provides inspectors with the necessary background to become certified.

    c. **Detail oriented.** Inspectors thoroughly examine many different construction activities. Therefore, they must pay close attention to detail so as to not overlook any items that need to be checked.

    d. **Mechanical knowledge.** Inspectors use a variety of testing equipment as they check complex systems. In order to perform tests properly, they also must have detailed knowledge of how the systems operate.

    e. **Physical stamina.** Inspectors are constantly on their feet and often climb and crawl through attics and other tight spaces. As a result, they should be somewhat physically fit.

10. **Union/non-union:** Although there is not a construction inspectors' union, there are many state associations for building inspectors and the ICC, a nonprofit association that provides a wide range of building safety solutions including product evaluation, accreditation, certification, codification and training. Unions can provide health insurance, job security, pensions, and representation.

11. **Compensation:** In 2019, median pay for construction and building inspectors was $60,710 annually, or $29.19 per hour, not including benefits such as health insurance and retirement. Starting annual pay is closer to $36,440. Top earners earn $98,820 annually or more. Note that generally, union jobs pay higher than non-union jobs, and cities pay higher than rural areas. In 2018, there were 117,300 construction and building inspector jobs in the U.S.

12. **Employment outlook:** Employment of construction and building inspectors is projected to grow 7 percent from 2018 to 2028, faster than the average for all occupations. Public interest in safety and the desire to improve the quality of construction should continue to create demand for inspectors. Certified construction and building inspectors who can perform a variety of inspections should have the best job opportunities.

See also: **Trade 4:** Carpenter
Trade 8: Electrician
Trade 19: Plumber, Pipefitter, and Steamfitter
Trade 22: Surveyor

TRADE 6

Diesel Technician

Do you like working on big trucks? Diesel technicians work on all aspects of trucks, buses, bulldozers, and some passenger cars and pickups.

1. **History of diesel technicians:** In the 1890s, Rudolf Diesel invented an efficient, compression ignition, internal combustion engine that now bears his name. (In gasoline engines, fuel is mixed with air, compressed by pistons and ignited by sparks from spark plugs. In a diesel engine, however, the air is compressed first, and then the fuel is injected.] Early diesel engines were large and operated at low speeds due to the limitations of their compressed air-assisted fuel injection systems. High-speed diesel engines were introduced in the 1920s for commercial vehicle applications and in the 1930s for passenger cars.

2. **What diesel technicians do:** Diesel technicians and mechanics often inspect vehicle brake systems, steering mechanisms, transmissions, and engines. They read and interpret diagnostic tests and repair or replace malfunctioning components of mechanical or electrical equipment. They also perform basic care and maintenance including oil changes, fluid level checks, and tire rotation. Diesel engines power trucks, buses, heavy equipment such as bulldozers, commercial boats, and some passenger vehicles and pickups.

3. **Work environment:** Diesel technicians often work in crowded, noisy shops. They may also make roadside repairs. Diesel service

technicians and mechanics often lift heavy parts and tools, handle greasy or dirty equipment, and work in uncomfortable positions. Most diesel technicians work full time. If a shop offers 24-hour repair, night, weekend, and holiday work may be required.

4. **Education needed**: Most employers require a high school diploma or equivalent. High school or college-level courses in automotive repair, electronics, and mathematics provide a strong educational background for a career as a diesel technician. Some employers prefer to hire workers with college education in diesel engine repair. Many community colleges and trade and vocational schools offer certificate or degree programs in diesel engine repair. These degree programs mix classroom instruction with hands-on training and include learning the basics of diesel technology, repair techniques and equipment, and practical exercises. Students also learn how to interpret technical manuals and electronic diagnostic reports.

5. **Training needed:** Diesel technicians are typically trained extensively on the job. Trainees are assigned basic tasks, such as cleaning parts, checking fuel and oil levels, and driving vehicles in and out of the shop. After learning routine maintenance and repair tasks, trainees move on to more complicated subjects, such as vehicle diagnostics. This process can take from 3 to 4 years, at which point a trainee is usually considered a journey-level diesel technician.

6. **License/certification**: Certification from the National Institute for Automotive Service Excellence (ASE) is the standard credential for diesel and other automotive service technicians and mechanics. Although not required, this certification demonstrates a diesel technician's competence and experience to potential employers and clients, and often brings higher salary. Diesel technicians may be certified in specific repair areas, such as drivetrains, electronic systems, and preventive maintenance and inspection. To earn certification, technicians must have 2 years of work experience and pass ASE exams. To remain certified,

diesel technicians must pass a recertification exam every 5 years. Some diesel technicians are required to have a commercial driver's license so that they may test-drive buses and large trucks.

7. **Competencies needed**: The following are required competencies for diesel technicians.

   a. **Customer-service skills**. Diesel technicians frequently discuss automotive problems and necessary repairs with their customers. They must be courteous, good listeners, and ready to answer customers' questions.

   b. **Detail oriented**. Diesel technicians must be aware of small details when inspecting or repairing engines and components, because mechanical and electronic malfunctions are often due to misalignments and other easy-to-miss causes.

   c. **Dexterity**. Mechanics need a steady hand and good hand–eye coordination for many tasks, such as disassembling engine parts, connecting or attaching components, and using hand tools.

   d. **Mechanical skills**. Diesel technicians must be familiar with engine components and systems and know how they interact with each other. They often disassemble major parts for repairs, and they must be able to put them back together properly.

   e. **Organizational skills**. Diesel technicians must keep workspaces clean and organized in order to maintain safety and accountability for parts.

   f. **Physical strength**. Diesel technicians often lift heavy parts and tools, such as exhaust system components and pneumatic wrenches.

   g. **Troubleshooting skills**. Diesel technicians use diagnostic equipment on engine systems and components in order to identify and fix problems in mechanical and electronic systems. They must be familiar with electronic control systems and the appropriate tools needed to fix and maintain them.

8. **Union/non-union:** Automotive workers are frequently represented by the International Association of Machinists (IAM) and Aerospace Workers. Unions can provide health insurance, job security, pensions, and representation.

9. **Compensation:** In 2019, median pay for diesel service technicians was $48,500 annually, or $23.32 per hour, not including benefits such as health insurance and retirement. Starting annual pay is closer to $31,990. Top earners earn $74,090 annually or more. Note that generally, union jobs pay higher than non-union jobs, and cities pay higher than rural areas. In 2018, there were 285,300 diesel service technician jobs in the U.S.

10. **Employment outlook:** Employment of diesel service technicians is projected to grow 5 percent from 2018 to 2028, about as fast as all occupations.

See also: **Trade 12:** Heavy Equipment Operator
Trade 24: Tractor Trailer Truck Driver

TRADE 7

Drafter

The typical image of a drafter is someone bent over a drafting table, using pencils and rulers to create a precise drawing of a building. The work is similar today, except now drafters use computers to create designs.

1. **History of draftsmanship:** The word *draftsman*, or *draughtsman*, dates back to the 1660s. In the past, drafters sat at drawing boards and used pencils, pens, compasses, protractors, triangles, and other drafting devices to prepare a drawing by hand. In the 1980s and 1990s, board drawings were replaced by newly developed computer-aided design (CAD) systems that could produce technical drawings at a faster pace. *Draughtsman* evolved to *draftsman/draftswoman* and now to the neutral *drafter*.

2. **What draftsmen/draftswomen do:** Drafters typically design plans using computer-aided design (CAD) software, work from rough sketches and specifications cr-eated by engineers and architects, design products with engineering and manufacturing techniques, add details to architectural plans from their knowledge of building techniques, and specify dimensions, materials, and procedures for new products. Specialties include:

3. **Architectural drafters**, who draw architectural and structural features of buildings for construction projects

4. **Civil drafters**, who prepare maps used in construction and civil engineering projects, such as highways, bridges, and flood-control projects

5. **Electrical drafters,** who prepare wiring diagrams that construction workers use to install and repair electrical equipment

6. **Electronics drafters,** who produce wiring diagrams, assembly diagrams for circuit boards, and layout drawings used in manufacturing and in installing and repairing electronic devices and components.

7. **Mechanical drafters,** who prepare layouts that show the details for a wide variety of machinery and mechanical tools and devices, such as medical equipment.

8. **Work environment:** Although drafters spend much of their time working on computers in an office, some may visit jobsites in order to collaborate with architects and engineers. Drafters are considered skilled assistants to architects and engineers.

9. **Education needed:** High school courses in English, mathematics, science, electronics, computer technology, drafting and design, visual arts, and computer graphics are useful for Millennials considering a drafting career. Drafters typically need an associate of applied science in drafting or a related degree from a community college or technical school. Some drafters prepare for the occupation by earning a certificate or diploma.

10. **Training needed:** Programs in drafting may include instruction in design fundamentals, sketching, and CAD software. It generally takes about 2 years of full-time education to earn an associate's degree. Certificate and diploma programs vary in length but usually may be completed in less time. Students frequently specialize in a particular type of drafting, such as mechanical or architectural drafting. High school students may begin preparing by taking classes in mathematics, science, computer technology, design, computer graphics, and, where available, drafting.

11. **License/certification:** The American Design Drafting Association (ADDA) offers certification for drafters. Although not mandatory, certification demonstrates competence and

knowledge of nationally recognized practices. Certifications are offered for several specialties, including architectural, civil, and mechanical drafting.

12. **Competencies needed**: The following are required competencies for drafters.

   a. **Creativity.** Drafters must be able to turn plans and ideas into technical drawings of buildings, tools, and systems.

   b. **Detail oriented.** Drafters must take care that the plans they convert are technically accurate according to the outlined specifications.

   c. **Interpersonal skills.** Drafters work closely with architects, engineers, and other designers to make sure that final plans are accurate. This requires the ability to communicate effectively and work well with others.

   d. **Math skills.** Drafters work on technical drawings. They may be required to calculate angles, weights, costs, and other values.

   e. **Technical skills.** Drafters in all specialties must be able to use computer software, such as CAD, and work with database tools, such as building information modeling (BIM).

   f. **Time-management skills.** Drafters often work under deadline. As a result, they must work efficiently to produce the required output according to set schedules.

13. **Union/non-union:** Although the ADDA is a trade association that represents drafters' interests, there is not currently a union for drafters.

14. **Compensation**: In 2019, median pay for drafters was $56,830 annually, or $27.32 per hour per hour, not including benefits such as health insurance and retirement. Starting annual pay is closer to $35,920. Top earners earn $87,720 annually or more.

Note that generally, union jobs pay higher than non-union jobs, and cities pay higher than rural areas. In 2018, there were 199,900 drafter jobs in the U.S.

15. **Employment outlook**: Overall employment of drafters is projected to grow 7 percent from 2016 to 2026. Employment growth will vary by specialty. Growth in the engineering services and construction industries is expected to account for most new jobs for drafters.

See also: **Trade 22:** Surveyor
Trade 23: Telecommunication Technician

TRADE 8

Electrician

What would we do without electricity? From lights to outlets to surge protection to fans, and many varieties of new construction and commercial electrical needs, electricians are essential to every kind of indoor living and working!

1. **History of electricians:** The first electrical streetlight was installed by the first electricians in Los Angeles in 1875. In the 1880s, the first power stations and transformers were created. Expositions and world fairs in the late 1800s became popular places to display new advances in electricity. Electricians were hired to build and operate the new findings at these expos, and then throughout all aspects of electricity in homes, businesses, and outside.

2. **What electricians do:** Electricians install, maintain, and repair electrical power, communications, lighting, and control systems.

3. **Work environment:** Electricians may work in residential or business settings. Almost all electricians work full time. Work schedules may include evenings and weekends. Overtime is common; overtime may pay more for union members than for non-union members.

4. **Education needed:** A high school diploma or equivalent is required to become an electrician.

5. **Training needed:** Some electricians start out by attending a technical school. Many technical schools offer programs related

to circuitry, safety practices, and basic electrical information. Graduates of these programs usually receive credit toward their apprenticeship.

6. **Apprenticeships:** Most electricians learn their trade in a 4- or 5-year apprenticeship program. For each year of the program, apprentices typically receive 2,000 hours of paid on-the-job training as well as some technical instruction. Workers who gained electrical experience in the military or in the construction industry may qualify for a shortened apprenticeship based on their experience and testing. Technical instruction for apprentices includes electrical theory, blueprint reading, mathematics, electrical code requirements, and safety and first-aid practices. They may also receive specialized training related to soldering, communications, fire alarm systems, and elevators. Several groups, including unions and contractor associations, sponsor apprenticeship programs. Apprenticeship requirements vary by state and locality. Some electrical contractors have their own training programs, which are not recognized apprenticeship programs but include both technical and on-the-job training. Although most workers enter apprenticeships directly, some electricians enter apprenticeship programs after working as a helper. The Home Builders Institute (HBI) offers a pre-apprenticeship certificate training (PACT) program for eight construction trades, including electricians.

7. **After apprenticeship:** After completing an apprenticeship program, electricians are considered journey workers and may perform duties on their own, subject to local or state licensing requirements.

8. **License/certification:** Most states require electricians to pass a test and be licensed. Requirements vary by state. Many of the requirements can be found on the National Electrical Contractors Association (NECA) website. The tests have questions related to the National Electrical Code (NEC) and state

and local electrical codes, all of which set standards for the safe installation of electrical wiring and equipment. Electricians may be required to take continuing education courses in order to maintain their licenses. These courses are usually related to safety practices, changes to the electrical code, and training from manufacturers in specific products. Electricians may obtain additional certifications, which demonstrate competency in areas such as solar photovoltaic, electrical generating, or lighting systems. Electricians may be required to have a driver's license.

9. **Competencies needed**: The following are required competencies for electricians.

   a. **Color vision.** Electricians must be able to identify electrical wires by color.

   b. **Critical-thinking skills.** Electricians perform tests and use the results to diagnose problems. For example, when an outlet is not working, they may use a multimeter to check the voltage, amperage, or resistance in order to determine the best course of action.

   c. **Customer-service skills.** Electricians interact with people on a regular basis. They should be friendly and be able to address customers' questions.

   d. **Physical stamina.** Electricians often need to move around all day while running wire and connecting fixtures to the wire.

   e. **Physical strength.** Electricians need to be strong enough to move heavy components, which may weigh up to 50 pounds.

   f. **Troubleshooting skills.** Electricians find, diagnose, and repair problems. For example, if a motor stops working, they perform tests to determine the cause of its failure and then, depending on the results, fix or replace the motor.

10. **Union/non-union:** The International Brotherhood of Electrical Workers (IBEW) is the largest organization of electrical workers in North America. Their 750,000 members work in nearly every setting, including construction sites, power plants, factories, offices, shipyards, TV studios and rail yards. Unions can provide health insurance, job security, pensions, and representation.

11. **Compensation:** In 2019, median pay for median pay for electricians was $56,180 annually, or $27.01 per hour, not including benefits such as health insurance and retirement. Starting annual pay is closer to $33,410. Top earners earn $96,580 annually or more. Note that generally, union jobs pay higher than non-union jobs, and cities pay higher than rural areas. In 2018, there were 715,400 electrician jobs in the U.S.

12. **Employment outlook:** Employment of electricians is projected to grow 10 percent from 2018 to 2028, faster than the average for all occupations.

See also: **Trade 4:** Carpenter
Trade 13: Ironworker
Trade 19: Plumber, Pipefitter and Steamfitter
Trade 25: Construction Manager/Project Manager

TRADE 9

Flooring Installer/Tile and Marble Setter

Do you ever think about the floor you walk on? Flooring installers create magic at our feet, whether tile, wood, vinyl, carpet, or any other kind of flooring that keeps our feet happy and brings together our home or workplace.

1. **History of flooring installers and tile and marble setters:** About 5,000 years ago, Egyptians began to create stone and brick floors. At first utilitarian, over time they became works of art with mosaic patterns made of tiles. Wood flooring was created in the early years of the U.S. when colonists cut old-growth forests into rough planks that were nailed to floor joists. Tongue-in-groove and parquet wood flooring followed. The creation of the power loom in the early 18th century led to carpeting, which was originally hand-sewn from strips and tacked down at the edges of a room. Wall to wall carpet dates back to the late 19th century.

2. **What flooring installers and tile and marble setters do:** Flooring installers and tile and marble setters lay and finish carpet, wood, vinyl, and tile.

3. **Work environment:** Installing flooring, tile, and marble is physically demanding, with workers spending much of their time reaching, bending, and kneeling. Those employed in commercial settings may work evenings and weekends.

4. **Education needed**: There are no specific education requirements for someone to become a flooring installer or tile and
   marble setter. A high school diploma or equivalent is preferred
   for those entering an apprenticeship program. High school art,
   math, and vocational courses are considered helpful for flooring
   installers and tile and marble setters.

5. **Training needed**: Flooring installers and tile and marble setters
   typically learn their duties through on-the-job training, working
   with experienced installers. Although workers may enter training directly, many start out as helpers. New workers usually start
   by performing simple tasks, such as moving materials. As they
   gain experience, they are given more complex tasks, such as
   cutting carpet. Some tile installer helpers become tile finishers
   before becoming tile installers.

6. **Apprenticeship**: Some flooring installers and tile and marble
   setters learn their trade through a 2- to 4-year apprenticeship.
   This instruction may include mathematics, building code requirements, safety and first-aid practices, and blueprint reading.

7. **After apprenticeship**: After completing an apprenticeship
   program, flooring installers and tile and marble setters are considered to be journey workers and may perform duties on their own.

8. **License/certification**: Several organizations and groups offer
   certifications for floor and tile installers. Although certification is
   not required, it demonstrates that a flooring installer and tile and
   marble setter has specific mastery skills to do a job. The Ceramic
   Tile Education Foundation (CTEF) offers the Certified Tile
   Installer (CTI) certification for workers with 2 or more years of
   experience as a tile installer. Applicants are required to complete a
   written test and a hands-on performance evaluation. Several groups,
   including the CTEF, the International Masonry Institute (IMI),
   the International Union of Bricklayers and Allied Craftworkers
   (BAC), the National Tile Contractors Association (NTCA), the
   Tile Contractors' Association of America (TCAA), and the Tile

Council of North America (TCNA) have created the Advanced Certifications for Tile Installers (ACT) program. Certification requirements include passing both an exam and a field test. Workers must also have either completed a qualified apprenticeship program or earned the CTI certification to qualify for testing. The National Wood Flooring Association (NWFA) has a voluntary certification for floor sanders and finishers. Sanders and finishers must have 2 years of experience and must have completed NWFA-approved training. Applicants are also required to complete written and performance tests. The International Certified Floorcovering Installers Association (CFI) offers certification for flooring and tile installers. Installers need 2 years of experience before they can take the written test and a hands-on performance evaluation. The International Standards & Training Alliance (INSTALL) offers a comprehensive flooring certification program for flooring and tile installers. INSTALL certification requires both classroom and hands-on training and covers all major types of flooring.

9. **Competencies needed**: The following are required competencies for flooring installers and tile and marble setters.

   a. **Color vision.** Flooring installers and tile and marble setters often determine small color variations. Because tile patterns may include many different colors, workers must be able to distinguish among colors and among patterns for the best-looking finish.

   b. **Customer-service skills.** Flooring installers and tile and marble setters commonly work in customers' homes. Therefore, workers must be courteous and considerate of a customer's property while completing tasks.

   c. **Detail oriented.** Flooring installers and tile and marble setters need to plan and lay out materials. Some carpet patterns can be highly detailed and artistic, so workers must ensure that the patterns are properly and accurately aligned.

d. **Math skills.** Flooring installers and tile and marble setters use measurement-related math skills on every job. Besides measuring the area to be covered, workers must calculate the number of carpet tiles needed to cover that area.

e. **Physical stamina.** Flooring installers and tile and marble setters must have the endurance to stand or kneel for many hours. Workers need to spread adhesives quickly and place tile on floors before the adhesives harden.

f. **Physical strength.** Flooring installers and tile and marble setters lift and carry heavy materials. Workers must be strong enough to lift, carry, and set heavy pieces of marble into position.

10. **Union/non-union:** Flooring installers and tile and marble setters can belong to the BAC or the International Union of Painters and Allied Trades (IUPAT). Unions can provide health insurance, job security, pensions, and representation.

11. **Compensation:** In 2019, median pay for median pay for flooring installers and tile and marble setters was $42,050 annually, or $20.22 per hour, not including benefits such as health insurance and retirement. Starting annual pay is closer to $25,780. Top earners earn $74,630 annually or more. Note that generally, union jobs pay higher than non-union jobs, and cities pay higher than rural areas. In 2018, there were 119,600 flooring installers and tile and marble setters in the U.S.

12. **Employment outlook:** Employment of flooring installers and tile and marble setters is projected to grow 11 percent from 2018 to 2028, much faster than the average for all occupations.

See also: Trade 2: Bricklayer/mason
Trade 11: Heating, Ventilation, Air-Conditioning, and Refrigeration Technician
Trade 18: Plasterer/Stucco Mason

TRADE 10

Glazier

As a child, I was always delighted when I saw a glazier truck with its big sheets of glass lined up so perfectly and carefully! Glaziers take care of all the glass in our lives, focusing on windows, skylights, and the framing that supports them.

1. **History of glaziers:** Glass has been around since the Stone Age, but the word *glazier* was derived from the Old English word *glaes,* which described a person who manufactured glass objects. Glass in its natural form is a bluish-green, which is caused by iron impurities from the sand used to make it. Glaziers change the color of glass by adding metallic compounds and mineral oxides to heated, liquid glass. Egyptians created the technique of changing the color of glass and glassblowing to create different glass objects such as jars and bottles, and Germans in the 11th century created new ways to make sheet glass.

2. **What glaziers do:** Glaziers design, cut, install, and repair glass in windows, skylights, and other fixtures in storefronts and buildings.

3. **Work environment:** The work of glaziers is physically demanding. They may experience cuts from tools and glass, falls from ladders and scaffolding, and exposure to solvents. Most work full time.

4. **Education needed:** Most programs require apprentices to have a high school diploma or equivalent and be at least 18 years old.

5. **Training/Apprenticeship:** Glaziers typically learn their trade through a 4-year apprenticeship or on-the-job training. On the job, they learn to use the tools and equipment of the trade; handle, measure, cut, and install glass and metal framing; cut and fit moldings; and install and balance glass doors. Technical training includes learning different installation techniques, as well as basic mathematics, blueprint reading and sketching, general construction techniques, safety practices, and first aid. A few groups sponsor apprenticeship programs, including several union and contractor associations.

6. **After apprenticeship:** After completing an apprenticeship program, glaziers are considered to be journey workers who may do tasks on their own.

7. **License/certification:** Some states may require glaziers to have a license; check with your state for more information. Licensure requirements typically include passing a test and possessing a combination of education and work experience.

8. **Competencies needed:** The following are required competencies for glaziers.

    a. **Balance.** Glaziers need a good sense of balance while handling large panes of glass or while working on ladders or scaffolds.

    b. **Communication.** Glaziers need to be able to communicate effectively with other team members and with customers to ensure the work is done precisely and on time.

    c. **Hand–eye coordination.** Glaziers must be able to cut glass precisely. As a result, a steady hand is needed to cut the correct size and shape in the field.

    d. **Physical stamina.** Glaziers work on their feet and move heavy pieces of glass most of the day. They need to be able to hold glass in place until it can be fully secured.

e. **Physical strength.** Glaziers must often lift heavy pieces of glass for hanging. Physical strength, therefore, is important for the occupation.

9. **Union/non-union:** Glaziers can belong to the Glaziers, Architectural Metal and Glassworkers union or to the International Union of Painters and Allied Trades (IUPAT). Unions can provide health insurance, job security, pensions, and representation.

10. **Compensation**: In 2019, median pay for median pay for glaziers was $44,630 annually, or $21.46 per hour, not including benefits such as health insurance and retirement. Starting annual pay is closer to $27,860. Top earners earn $83,780 annually or more. Note that generally, union jobs pay higher than non-union jobs, and cities pay higher than rural areas. In 2018, there were 53,500 glaziers in the U.S.

11. **Employment outlook**: much faster than the average for all occupations.

See also: Trade 3: Cabinetmaker/Woodworker
Trade 9: Flooring Installer/Tile and Marble Setter
Trade 21: Sheet Metal Worker

TRADE 11

Heating, Ventilation, Air-Conditioning, and Refrigeration Technician

Heating, air-conditioning, and refrigeration are truly world wonders. If you doubt it, consider what it's like to live in a climate that's very hot or very cold, or in a place where you can't store food for any amount of time. HVACR technicians keep us comfortable and fed!

1. **History of HVACR technicians:** In 1902, Willis Carrier invented a machine called the Apparatus for Treating Air to keep paper from being ruined by humidity. It functioned by blowing air over cold coils to control temperature and humidity. In 1931, window air-conditioning units were created, and 1939 saw the first air-conditioned car. In the 1970s, central air, including a condenser, coils, and fan, revolutionized air-conditioning—and greatly increased demand for HVACR technicians.

2. **What HVACR technicians do:** Heating, air-conditioning, and refrigeration technicians install and repair heating, ventilation, cooling, and refrigeration systems.

3. **Work environment:** HVACR technicians work mostly in homes, schools, hospitals, office buildings, or factories. Their worksites may be very hot or cold because the heating and cooling systems

they must repair may not be working properly and because some parts of these systems are located outdoors. Working in cramped spaces and during irregular hours is common.

4. **Education needed**: High school students interested in becoming an HVACR technician should take courses in vocational education, math, and physics. Knowledge of plumbing or electrical work and a basic understanding of electronics is also helpful.

5. **Training needed:** Many HVACR technicians receive postsecondary instruction from technical and trade schools or community colleges that offer programs in heating, air-conditioning, and refrigeration. These programs generally last from 6 months to 2 years and lead to a certificate or an associate's degree. Because HVACR systems have become increasingly complex, employers generally prefer applicants with postsecondary education or those who have completed an apprenticeship.

6. **Apprenticeships:** New HVACR technicians typically begin by working alongside experienced technicians. At first, they perform basic tasks such as insulating refrigerant lines or cleaning furnaces. In time, they move on to more difficult tasks, including cutting and soldering pipes or checking electrical circuits. Some technicians receive their training through an apprenticeship. Apprenticeship programs usually last 3 to 5 years. Over the course of the apprenticeship, technicians learn safety practices, blueprint reading, and how to use tools. They also learn about the numerous systems that heat and cool buildings. Several groups, including unions and contractor associations, sponsor apprenticeship programs. Apprenticeship requirements vary by state and locality.

7. **License/certification**: Some states and localities require HVACR technicians to be licensed. The U.S. Environmental Protection Agency (EPA) requires all technicians who buy, handle, or work with refrigerants to be certified in proper refrigerant handling. Many trade schools, unions, and employer

associations offer training programs designed to prepare students for the EPA certification exam. Workers may need to pass a background check prior to being hired.

8. **Competencies needed:** The following are required competencies for HVACR technicians.

   a. **Customer-service skills.** HVACR technicians often work in customers' homes or business offices, so it is important that they be friendly, polite, and punctual. Repair technicians sometimes deal with unhappy customers whose heating or air-conditioning is not working.

   b. **Detail oriented.** HVACR technicians must carefully maintain records of all work performed. The records must include the nature of the work performed and the time it took, as well as a list of specific parts and equipment that were used.

   c. **Math skills.** HVACR technicians need to calculate the correct load requirements to ensure that the HVACR equipment properly heats or cools the space required.

   d. **Mechanical skills.** HVACR technicians install and work on complicated climate-control systems, so they must understand the HVAC components and be able to properly assemble, disassemble, and, if needed, program them.

   e. **Physical stamina.** HVACR technicians may spend many hours walking and standing. The constant physical activity can be tiring.

   f. **Physical strength.** HVACR technicians may have to lift and support heavy equipment and components, often without help.

   g. **Time-management skills.** HVACR technicians frequently have a set number of daily maintenance calls. They should be able to keep a schedule and complete all necessary repairs or tasks.

h. **Troubleshooting skills.** HVACR technicians must be able to identify problems on malfunctioning heating, air-conditioning, and refrigeration systems and then determine the best way to repair them.

9. **Union/non-union:** HVACR technicians often join the United Association of Journeymen and Apprentices of the Plumbing and Pipe Fitting Industry of the United States, Canada (UA)— typically referred to as United Association—or the International Association of Sheet Metal, Air, Rail and Transportation Workers (SMART).

10. **Compensation:** In 2019, median pay for median pay for HVACR technicians was $48,730 annually, or $23.43 per hour, not including benefits such as health insurance and retirement. Starting annual pay is closer to $30,610. Top earners earn $77,920 annually or more. Note that generally, union jobs pay higher than non-union jobs, and cities pay higher than rural areas. In 2018, there were 367,900 HVACR technicians in the U.S.

11. **Employment outlook:** Employment of HVACR technicians is projected to grow 13 percent from 2018 to 2028, much faster than the average for all occupations.

See also: **Trade 1:** Boilermaker
Trade 8: Electrician
Trade 15: Machinist/Tool and Die Maker
Trade 21: Sheet Metal Worker

TRADE 12

Heavy Equipment Operator

Many of us were fascinated as kids with heavy equipment such as huge trucks, tractors, harvesters, excavators, backhoes, cranes, or bulldozers, but only a select few of us get to operate those for a living. It can be a dirty job, but heavy equipment operators are literally moving the earth (and many other things as well).

1. **History of heavy equipment operators:** Until nearly the twentieth century, the primary method for moving earth was the shovel, supplemented with power from humans and animals. Portable steam power made new engines possible, and traction engines, internal combustion, kerosene, and ethanol led to new developments in engines. Today, diesel engines dominate in heavy equipment operations.

2. **What heavy equipment operators do:** Heavy equipment operators drive, maneuver, or control the heavy machinery used to construct roads, buildings and other structures.

3. **Work environment:** Heavy equipment operators work in nearly every weather condition. They often get dirty, greasy, muddy, or dusty. The majority of operators work full time, and some operators have irregular work schedules. Some construction projects, especially road building, are done at night.

4. **Education needed:** A high school diploma or equivalent is

required for most jobs. Vocational training and math courses are useful, and a course in auto mechanics can be helpful because workers often perform maintenance on their equipment. Learning at vocational schools may be beneficial in finding a job. Schools may specialize in a particular brand or type of construction equipment. Some schools incorporate sophisticated simulator training into their courses, allowing beginners to familiarize themselves with the equipment in a virtual environment before operating real machines.

5. **Training/Apprenticeship:** Many workers learn their jobs by operating light equipment under the guidance of an experienced operator. Later, they may operate heavier equipment, such as bulldozers. Some construction equipment with computerized controls requires greater skill to operate. Operators of such equipment may need more training and some understanding of electronics. Other workers learn their trade through a 3- or 4-year apprenticeship. For each year of the program, apprentices must have at least 144 hours of technical instruction and 2,000 hours of paid on-the-job training. On the job, apprentices learn to maintain equipment, operate machinery, and use technology, such as Global Positioning System (GPS) devices. In the classroom, apprentices learn operating procedures for equipment, safety practices, and first aid, as well as how to read grading plans. A few groups, including unions and contractor associations, sponsor apprenticeship programs. The basic qualifications for entering an apprenticeship program are as follows: Minimum age of 18, high school education or equivalent, physically able to do the work, and valid driver's license.

6. **After apprenticeship:** After completing an apprenticeship program, heavy equipment operators are considered journey workers and perform tasks with less guidance.

7. **License/certification:** Heavy equipment operators often need a commercial driver's license (CDL) to haul their equipment to

various jobsites. State laws governing CDLs vary. A few states have special licenses for operators of backhoes, loaders, and bulldozers. Currently, 17 states require pile-driver operators to have a crane license because similar operational concerns apply to both pile-drivers and cranes. In addition, the cities of Chicago, Cincinnati, New Orleans, New York, Omaha, Philadelphia, and Washington, DC require special crane licensure.

8. **Competencies needed:** The following are required competencies for heavy equipment operators.

   a. **Hand-eye-foot coordination.** Heavy equipment operators should have steady hands and feet to guide and control heavy machinery precisely, sometimes in tight spaces.

   b. **Mechanical skills.** Heavy equipment operators often perform basic maintenance on the equipment they operate. As a result, they should be familiar with hand and power tools and standard equipment care.

   c. **Physical strength.** Heavy equipment operators may be required to lift more than 50 pounds as part of their duties.

   d. **Unafraid of heights.** Heavy equipment operators may work at great heights. For example, pile-driver operators may need to service the pulleys located at the top of the pile-driver's tower, which may be several stories tall.

9. **Union/non-union:** Heavy equipment operators often join the International Union of Operating Engineers (IUOE). Unions can provide health insurance, job security, pensions, and representation.

10. **Compensation:** In 2019, median pay for median pay for heavy equipment operators was $48,160 annually, or $23.16 per hour, not including benefits such as health insurance and retirement. Starting annual pay is closer to $31,780. Top earners earn $84,650 annually or more. Note that generally, union jobs pay

higher than non-union jobs, and cities pay higher than rural areas. In 2018, there were 453,200 heavy equipment operators in the U.S.

11. **Employment outlook**: Employment of heavy equipment operators is projected to grow 10 percent from 2018 to 2028, faster than the average for all occupations.

See also: **Trade 6:** Diesel Technician
Trade 13: Ironworker
Trade 15: Machinist/Tool and Die Maker
Trade 21: Sheet Metal Worker

TRADE 13

Ironworker

As skyscrapers began to rise in New York in the 1920s, Mohawks from local tribes became the premier ironworkers, able to walk steel beams 30 stories or more above the city. Ironworkers were captured in the iconic photo, "Lunch Atop a Skyscraper," depicting 11 ironworkers having a leisurely break 70 stories up. Interested? Read on!

1. **History of ironworking:** The earliest known ironwork are beads from Egypt dating back to 3500 B.C. and made from meteoric iron. The earliest use of smelted iron dates back to Mesopotamia. The first use of conventional smelting and purification techniques that modern society labels as true ironworking dates back to the Hittites in around 2000 B.C. The Chinese were the first to use cast iron from the 6th century AD using it as support for pagodas and other buildings. In the 1880s, carpenters who built bridges became ironworkers.

2. **What ironworkers do:** Ironworkers work with iron to create buildings, reinforce structures, or create decorative iron constructions. Ironworkers may work on factories, steel mills, and utility plants. Ironworkers also perform all types of industrial maintenance. Three types of ironworkers are:

3. **Structural ironworker.** Structural ironworkers assemble cranes in order to lift the steel columns, beams, girders and trusses according to structural blueprints.

4. **Reinforcing ironworker.** A reinforcing bar (rebar) ironworker

works with reinforcing bars to make structures based on a certain design.

5. **Ornamental ironworker.** Ornamental ironworkers erect metal windows, stairways, catwalks, gratings, ladders, doors of all types, railings, fencing, gates, metal screens, elevator fronts, platforms, and entranceways by use of bolting or welding.

6. **Work environment:** Ironworkers perform physically demanding and dangerous work, often at great heights. Workers must wear safety equipment to reduce the risk of falls or other injuries.

7. **Education needed:** A high school diploma or equivalent is generally required to enter an apprenticeship. Workers learning through on-the-job training may not need a high school diploma or equivalent. Courses in math, as well as training in vocational subjects such as blueprint reading and welding, are useful.

8. **Apprenticeships:** Many ironworkers learn their trade through a 3- or 4-year apprenticeship. Sponsors of apprenticeship programs, nearly all of which teach both reinforcing and structural ironworking, include unions and contractor associations. For each year of the program, apprentices must have at least 144 hours of related technical instruction and 2,000 hours of paid on-the-job training. Ironworkers who complete an apprenticeship program are considered journey-level workers and may perform tasks without direct supervision. Other ironworkers receive on-the-job training that varies in length and is provided by their employer. On the job, apprentices and trainees learn to use the tools and equipment of the trade; handle, measure, cut, and lay rebar; and construct metal frameworks. They also learn about topics such as blueprint reading and sketching, general construction techniques, safety practices, and first aid.

9. License/certification: Certifications in welding, rigging, and crane signaling may make ironworkers more attractive to prospective employers. Several organizations provide certifications

for different aspects of the work. For example, the American Welding Society (AWS) offers welding certification, and several organizations offer rigging certifications, including the National Commission for the Certification of Crane Operators (NCCCO) and the National Center for Construction Education and Research (NCCER).

10. **Competencies needed**: The following are required competencies for ironworkers.

   a. **Ability to work at heights.** Ironworkers must not be afraid to work at great heights. For example, workers connecting girders during skyscraper construction may have to walk on narrow beams that are 50 stories or higher.

   b. **Balance.** Ironworkers often walk on narrow beams, so a good sense of balance is important to keep them from falling.

   c. **Critical thinking.** Ironworkers need to identify problems, monitor and assess potential risks, and evaluate the best courses of action. They must use logic and reasoning when finding alternatives so that they safely accomplish their tasks

   d. **Depth perception.** Ironworkers often signal crane operators who move beams and bundles of rebar, so they must be able to judge the distance between objects.

   e. **Hand-eye coordination.** Ironworkers must be able to tie rebar together quickly and precisely.

   f. **Physical stamina.** Ironworkers must have physical endurance because they spend many hours each day performing physically demanding tasks, such as moving rebar.

   g. **Physical strength.** Ironworkers must be strong enough to guide heavy beams into place and tighten bolts.

11. **Union/non-union:** Ironworkers typically join the International Association of Bridge, Structural, Ornamental, and Reinforcing

Iron Workers. Unions can provide health insurance, job security, pensions, and representation.

12. **Compensation**: In 2019, median pay for median pay for iron-workers was $53,650 annually, or $25.79 per hour, not including benefits such as health insurance and retirement. Starting annual pay is closer to $32,930. Top earners earn $89,790 annually or more. Note that generally, union jobs pay higher than non-union jobs, and cities pay higher than rural areas. In 2018, there were 98,600 ironworkers in the U.S.

13. **Employment outlook**: Employment of ironworkers is projected to grow 9 percent from 2018 to 2028, faster than the average of all occupations.

See also: **Trade 1:** Boilermaker
Trade 15: Machinist/Tool and Die Maker
Trade 21: Sheet Metal Worker

TRADE 14

Laborer

Laborers do physical work for wages. They unload, load, lift, carry, and hold construction materials on a jobsite. There is no doubt about it: this is physically demanding work.

1. **History of laborers:** Laborers are the valuable raw muscle supporting construction. Laborers manage demolition, paving, loading, utilities, piping, paving, and more. The 1st century B.C. engineer Vitruvius wrote that a good crew of laborers is just as valuable as any other aspect of construction.

2. **What laborers do:** Construction laborers and helpers perform many tasks that require physical labor on construction sites.

3. **Work environment:** Most construction laborers and helpers typically work full time and do physically demanding work. Some work at great heights or outdoors in all weather conditions. Construction laborers have one of the highest rates of injuries and illnesses of all occupations.

4. **Education needed:** Although formal education is not typically required for most laborer positions, helpers of electricians and helpers of pipelayers, plumbers, pipefitters, and steamfitters typically need a high school diploma. High school classes in mathematics, blueprint reading, welding, and other vocational subjects can be helpful.

5. **Training/Apprenticeship:** Construction laborers and helpers typically learn through on-the-job training after being hired by

a construction contractor. Workers usually learn by performing tasks under the guidance of experienced workers. Although the majority of construction laborers and helpers learn by assisting experienced workers, some construction laborers may opt for apprenticeship programs. These programs generally include 2 to 4 years of technical instruction and on-the-job training. The Laborers' International Union of North America (LIUNA) requires a combination of on-the-job training and related classroom instruction in such areas as signaling, blueprint reading, using proper tools and equipment, and following health and safety procedures. The remainder of the curriculum consists of specialized training in one of these eight areas: Building construction, Demolition and deconstruction, Environmental remediation, Road and utility construction, Tunneling, Masonry, Landscaping, and Pipeline construction.

6. **After apprenticeship:** After completing an apprenticeship program, apprentices are considered journey workers and perform tasks with less guidance.

7. **License/certification:** Laborers who remove hazardous materials (hazmat) must meet the federal and state requirements for hazardous materials removal workers. Depending on the work they do, laborers may need specific certifications, which may be attained through LIUNA. Rigging and scaffold building are commonly attained certifications. Certification can help workers prove that they have the knowledge to perform more complex tasks.

8. **Competencies needed:** The following are required competencies for laborers.

    a. **Color vision.** Construction laborers and helpers may need to be able to distinguish colors to do their job. For example, an electrician's helper must be able to distinguish different colors of wire to help the lead electrician.

b. **Math skills.** Construction laborers and some helpers need to perform basic math calculations while measuring on jobsites or assisting a surveying crew.

c. **Mechanical skills.** Construction laborers are frequently required to operate and maintain equipment, such as jackhammers.

d. **Physical stamina.** Construction laborers and helpers must have the endurance to perform strenuous tasks throughout the day. Highway laborers, for example, spend hours on their feet—often in hot temperatures—with few breaks.

e. **Physical strength.** Construction laborers and helpers must often lift heavy materials or equipment. For example, cement mason helpers must move cinder blocks, which typically weigh more than 40 pounds each.

9. **Union/non-union:** LIUNA represents laborers in North America for both public and private projects. Unions can provide health insurance, job security, pensions, and representation.

10. **Compensation:** In 2019, median pay for median pay for laborers was $36,000 annually, or $17.31 per hour, not including benefits such as health insurance and retirement. Starting annual pay is closer to $24,240. Top earners earn $64,100 annually or more. Note that generally, union jobs pay higher than non-union jobs, and cities pay higher than rural areas. In 2018, there were 453,200 heavy equipment operators in the U.S.

11. **Employment outlook:** Employment of laborers is projected to grow 11 percent from 2018 to 2028, much faster than the average for all occupations. Through experience and training, construction laborers and helpers can advance into positions that involve more complex tasks. For example, laborers may earn certifications in welding, erecting scaffolding, or finishing concrete, and then spend more time performing those activities.

Similarly, helpers sometimes move into construction craft occupations after gaining experience in the field. For example, experience as an electrician's helper may lead someone to becoming an apprentice electrician.

See also: **Trade 12:** Heavy Equipment Operator
Trade 21: Sheet metal worker
Trade 24: Tractor Trailer Truck Driver

TRADE 15

Machinist/Tool and Die Maker

Machinists do to metal what carpenters do to wood: they operate, disassemble, reassemble and repair metal tools and build new parts such as gears and shafts using various machine tools such as mills, lathes, grinders, and planers.

1. **History of machinists and tool and die makers**: In the 18th century, a machinist was simply someone who built or repaired machines. Typically, this work was done by hand, including forging and filing metal manually. By the middle of the 19th century, machine processes such as turning, boring, drilling, and planning was done by using lathes, milling machines, drill presses, or other mechanical methods. Since World War II, machining has expanded to include electron beam machining, photochemical machining, and ultrasonic machining in addition to traditional processes.

2. **What machinists and tool and die makers do**: Machinists and tool and die makers set up and operate machine tools to produce precision metal parts, instruments, and tools.

3. **Work environment**: Machinists and tool and die makers work in machine shops, toolrooms, and factories. Although many machinists and tool and die makers work full time during regular business hours, overtime, evening, and weekend work

may be common. Overtime may pay more for union members than for non-union members.

4. **Education needed**: Machinists typically have a high school diploma or equivalent, whereas tool and die makers may need to complete courses beyond high school. High school courses in math, blueprint reading, metalworking, and drafting are considered useful. Some community colleges and technical schools have 2-year programs that train students to become machinists or tool and die makers. These programs usually teach design and blueprint reading, the use of a variety of welding and cutting tools, and the programming and function of computer numerically controlled (CNC) machines.

5. **Training needed:** There are multiple ways for workers to gain competency on the job as a machinist or tool or die maker. One common way is through long-term on-the-job training, which lasts 1 year or longer. Trainees usually work 40 hours per week and take additional technical instruction during evenings. Trainees often begin as machine operators and gradually take on more difficult assignments. Machinists and tool and die makers must be experienced in using computers to work with computer-aided design/computer-aided manufacturing (CAD/CAM) technology, CNC machine tools, and computerized measuring machines. Some machinists become tool and die makers.

6. **Apprenticeships**: Some new workers may enter apprenticeship programs, which are typically sponsored by a manufacturer. Apprenticeship programs often consist of paid shop training and related technical instruction lasting several years. The technical instruction usually is provided in cooperation with local community colleges and vocational–technical schools. Workers typically enter into apprenticeships with a high school diploma or equivalent.

7. **License/certification**: A number of organizations and colleges offer certification programs. The Skills Certification System, for

example, is an industry-driven program that aims to align education pathways with career pathways. In addition, journey-level certification is available from state apprenticeship boards after the completion of an apprenticeship. Completing a certification program provides machinists and tool and die makers with better job opportunities and helps employers judge the abilities of new hires.

8. **Competencies needed:** The following are required competencies for machinists and tool and die makers.

   a. **Analytical skills.** Machinists and tool and die makers must understand technical blueprints, models, and specifications so that they can craft precision tools and metal parts.

   b. **Manual dexterity.** Machinists' and tool and die makers' work must be accurate. For example, machining parts may demand accuracy to within .0001 of an inch, a level of accuracy that requires workers' concentration and dexterity.

   c. **Math skills and computer application experience.** Workers must be experienced in using computers to work with CAD/CAM technology, CNC machine tools, and computerized measuring machines.

   d. **Mechanical skills.** Machinists and tool and die makers must operate milling machines, lathes, grinders, laser and water cutting machines, wire electrical discharge machines, and other machine tools.

   e. **Physical stamina.** Machinist and tool and die makers must stand for extended periods and perform repetitious movements.

   f. **Technical skills.** Machinists and tool and die makers must understand computerized measuring machines and metalworking processes, such as stock removal, chip control, and heat treating and plating.

9. **Union/non-union:** Machinists and tool and die makers typically join the International Association of Machinists and Aerospace Workers (IAM). Unions can provide health insurance, job security, pensions, and representation.

10. **Compensation:** In 2019, median pay for median pay for machinists and tool and die makers was $45,750 annually, or $21.99 per hour, not including benefits such as health insurance and retirement. Starting annual pay is closer to $27,940. Top earners earn $66,610 annually or more. Note that generally, union jobs pay higher than non-union jobs, and cities pay higher than rural areas. In 2018, there were 469,500 machinists and tool and die makers in the U.S.

11. **Employment outlook:** Employment of machinists and tool and die makers is projected to grow 11 percent from 2018 to 2028, much faster than the average for all occupations.

See also: Trade 1: Boilermaker
Trade 16: Millwright

TRADE 16

Millwright

Millwrights erect machinery. From an illustrious history creating mills that ground flour or created paper, millwrights now are industrial mechanics who install, maintain, and move machinery.

1. **History of millwrights:** The millwright profession dates back to the 12th century. As the name suggests, the original function of a millwright was the construction of wooden flour mills, sawmills, and paper mills powered by water or wind. Since the use of these structures originate in antiquity, millwrighting could arguably be considered one of the oldest engineering trades and the forerunner of modern mechanical engineering. Today, millwrights are high-precision tradespeople who install, dismantle, maintain, repair, reassemble, and move machinery in factories, power plants, and construction sites.

2. **What millwrights do:** Industrial machinery mechanics, machinery maintenance workers, and millwrights install, maintain, and repair factory equipment and other industrial machinery. This includes such tasks as leveling, aligning, and installing machinery on foundations or base plates, or setting, leveling, and aligning electric motors or other power sources such as turbines with the equipment.

3. **Work environment:** Most millwrights work full time in manufacturing facilities. However, they may be on call and work night or weekend shifts. Workers in this occupation must follow safety precautions and use protective equipment, such

as hardhats, safety glasses, and hearing protectors. Overtime is common; overtime may pay more for union members than for non-union members.

4. **Education needed**: Industrial machinery mechanics, machinery maintenance workers, and millwrights generally need at least a high school diploma or equivalent. Some mechanics and millwrights complete a 2-year associate's degree program in industrial maintenance. Industrial maintenance programs may include courses such as welding, mathematics, hydraulics, and pneumatics.

5. **Training needed:** Millwrights must have a good understanding of fluid mechanics (hydraulics and pneumatics) and all the components involved in these processes, such as valves, pumps, and compressors. They are also trained to work with a wide array of precision tools, such as calipers, micrometers, dial indicators, levels, gauge blocks, and optical and laser alignment tooling. Industrial machinery mechanics and machinery maintenance workers typically receive more than a year of on-the-job training. Industrial machinery mechanics and machinery maintenance workers learn how to perform routine tasks, such as setting up, cleaning, lubricating, and starting machinery. They also may be instructed in subjects such as shop mathematics, blueprint reading, proper hand tool use, welding, electronics, and computer programming. This training may be offered on the job by professional trainers hired by the employer or by representatives of equipment manufacturers.

6. **Apprenticeships:** Most millwrights learn their trade through a 3- or 4-year apprenticeship. For each year of the program, apprentices must have at least 144 hours of relevant technical instruction and up to 2,000 hours of paid on-the-job training. On the job, apprentices learn to set up, clean, lubricate, repair, and start machinery. During technical instruction, they are taught welding, mathematics, how to read blueprints, and machinery

troubleshooting. Many also receive computer training. The basic qualifications for entering an apprenticeship program are as follows: minimum age of 18, high school diploma or equivalent, and physically able to do the work. Employers, local unions, contractor associations, and the state labor department often sponsor apprenticeship programs.

7. **After apprenticeship:** After completing an apprenticeship program, millwrights are considered fully qualified and can usually perform tasks with less guidance.

8. **License/certification:** None required.

9. **Competencies needed:** The following are required competencies for millwrights.

   a. **Manual dexterity.** Industrial machinery mechanics, machinery maintenance workers, and millwrights must have a steady hand and good hand–eye coordination when handling very small parts.

   b. **Mechanical skills.** Industrial machinery mechanics, machinery maintenance workers, and millwrights use technical manuals and sophisticated diagnostic equipment to figure out why machines are not working. Workers must be able to reassemble large, complex machines after finishing a repair.

   c. **Troubleshooting skills.** Industrial machinery mechanics, machinery maintenance workers, and millwrights must observe, diagnose, and fix problems that a machine may be having.

10. **Union/non-union:** Millwrights must have 10 years of experience and seniority before being allowed to acquire one's journeyman card from the International Union. Within the Steelworkers Union, known as the United Steelworkers or USW, the largest Industrial Union in North America, there is also a mix of both

classroom and on-the-job training. Unions can provide health insurance, job security, pensions, and representation.

11. **Compensation**: In 2019, median pay for median pay for millwrights was $52,860 annually, or $25.41 per hour, not including benefits such as health insurance and retirement. Starting annual pay is closer to $35,030. Top earners earn $81,080 annually or more. Note that generally, union jobs pay higher than non-union jobs, and cities pay higher than rural areas. In 2018, there were 506,900 millwrights in the U.S.

12. **Employment outlook**: Employment of millwrights is projected to grow 5 percent from 2018 to 2028, about as fast as the average for all occupations.

See also: Trade 1: Boilermaker
Trade 11: Heating, Ventilation, Air-conditioning, and Refrigeration Technician
Trade 15: Machinist/Tool and Die Maker

TRADE 17

Painter

From artistic expression to covering walls, homes, and professional buildings, painting improves the appearance of buildings and protects them from damage by water, corrosion, insects and mold.

1. **History of painters:** Painting dates from pre-historic humans and has been identified in all cultures and continents. Paint itself was identified from cave paintings drawn with red or yellow ochre, hematite, manganese oxide, and charcoal made by humans as much as 40,000 years ago. Modern professional painting, such as painting homes and other buildings, dates to at least 2,000 years ago. In the late 1800s, Sherwin-Williams in the U.S. opened as a large paint-maker and invented a paint that could be used from the container without preparation. This made painting houses and other buildings much easier.

2. **What painters do:** Painting and coating workers paint and coat a wide range of products, often with the use of machines. Professional painters prepare and paint interior and exterior surfaces. They may work on residential or commercial properties. Professional painters often remove old paint, prime surfaces, choose materials, select and mix colors, apply paint, and clean up jobsites.

3. **Work environment:** Most painting and coating workers are employed full time. They frequently stand for long periods in specially ventilated areas.

4. **Education needed**: Painting and coating workers usually must have a high school diploma or equivalent. However, some employers hire workers who lack a high school diploma. Taking high school courses in automotive painting can be helpful. Some automotive painters attend a technical or vocational school where they receive hands-on training and learn the intricacies of mixing and applying different types of paint.

5. **Training needed:** Most painting and coating workers learn on the job after earning a high school diploma or equivalent. Although some worker training may last only a few days, most entry-level workers receive on-the-job training that may last from 1 month to several months. Workers who operate computer-controlled equipment may require additional training in computer programming.

6. **License/certification**: None required.

7. **Competencies needed**: The following are required competencies for painters.

   a. **Artistic ability.** Some painting and coating workers make elaborate or decorative designs. For example, some automotive painters specialize in making custom designs for vehicles.

   b. **Color vision.** Workers must be able to blend new paint colors in order to match existing colors on a surface.

   c. **Mechanical skills.** Workers must be able to operate and maintain sprayers that apply paints and coatings.

   d. **Physical stamina.** Some workers must stand at their station for extended periods. Continuous standing or activity can be tiring.

   e. **Physical strength.** Workers may need to lift heavy objects.

Some products that are painted or coated may weigh more than 50 pounds.

8. **Union/non-union:** Painters typically join the International Union of Painters and Allied Trades (IUPAT). Unions can provide health insurance, job security, pensions, and representation.

9. **Compensation:** In 2019, median pay for median pay for painters was $36,810 annually, or $17.70 per hour, not including benefits such as health insurance and retirement. Starting annual pay is closer to $24,520. Top earners earn $59,810 annually or more. Note that generally, union jobs pay higher than non-union jobs, and cities pay higher than rural areas. In 2018, there were 163,100 painters in the U.S.

10. **Employment outlook:** Employment of painters is projected to grow 2 percent from 2018 to 2028, slower than the average for all occupations.

See also: **Trade 3:** Cabinetmaker/Woodworker
Trade 7: Drafter
Trade 18: Plasterer/Stucco Mason
Trade 22: Surveyor

TRADE 18

Plasterer/ Stucco Mason

Plasterers are skilled craft workers who strengthen, soundproof, insulate, and fireproof buildings. Outside of buildings, they use cement plasters or stucco; indoors, plaster seals and beautifies walls.

1. **History of plasterers and stucco masons:** Early Greek architecture displays lime stucco and plaster with exquisite composition: white, fine and thin, often no thicker than an eggshell. Some of the plaster samples have weathered better than the stone to which it was applied. In its ornamental usage, plaster is an art form with many styles and types of enrichments. Plaster is a lining for partitions that divide room areas for privacy and diversified functional usages. As a fireproofing coating, plaster has been required by building codes and by law since the time of the Babylonians. Fireproofing of metal structural framing is a large market for the plastering industry today and provides a life/safety benefit to the public.

2. **What plasterers and stucco masons do:** Plasterers, stucco masons, and drywall and ceiling tile installers are specialty construction workers who build, apply, or fasten interior and exterior wallboards or wall coverings in residential, commercial, and other structures. Specifically, drywall and ceiling tile installers and tapers work indoors, installing wallboards to ceilings or to interior walls of buildings; plasterers and stucco masons, on the

other hand, work both indoors and outdoors—applying plaster to interior walls and cement or stucco to exterior walls. While most work is performed for functionality, such as fireproofing and sound dampening, some applications are intended purely for decorative purposes.

3. **Work environment:** As in many other construction trades, this work is physically demanding. Plasterers, stucco masons, and drywall and ceiling tile installers spend most of the day on their feet, either standing, bending, stretching, or kneeling. Some workers need to use stilts; others may have to lift and maneuver heavy, cumbersome materials, such as oversized wallboards. The work also can be dusty and dirty, irritating the skin, eyes, and lungs, unless protective masks, goggles, and gloves are used. Hazards include falls from ladders and scaffolds, and injuries from power tools and from working with sharp tools, such as utility knives. Most work indoors, except for the relatively few stucco masons who apply exterior finishes.

4. **Education needed:** A high school education is helpful, as are courses in basic math, mechanical drawing, and blueprint reading. The most common way to get a first job is to find an employer who will provide on-the-job training. Entry-level workers generally start as helpers, assisting more experienced workers. Employers may also send new employees to a trade or vocational school or community college to receive classroom training.

5. **Training needed:** Training is typically provided on the job through apprenticeships.

6. **Apprenticeships:** Some employers, particularly large nonresidential construction contractors with unionized workforces, offer employees formal apprenticeships. These programs combine on-the-job training with related classroom instruction—at least 144 hours of instruction each year for drywall and ceiling tile installers and tapers, and 166 hours for plasterers and stucco masons.

The length of the apprenticeship program, usually 3 to 4 years, varies with the apprentice's skill. Because the number of apprenticeship programs is limited, however, only a small proportion of these workers learn their trade this way. Helpers and apprentices start by carrying materials, lifting and cleaning up debris. They also learn to use the tools, machines, equipment, and materials of the trade. Within a few weeks, they learn to measure, cut, apply, and install materials. Eventually, they become fully experienced workers. At the end of their training, workers learn to estimate the cost of completing a job.

7. **License/certification**: Plasterers or stucco masons who complete apprenticeships registered with the federal or state government receive a journey worker certificate that is recognized nationwide.

8. **Competencies needed**: The following are required competencies for plasterers and stucco masons.

    a. **Artistic ability.** Artistic creativity is helpful for plasterers and stucco masons who apply decorative finishes.

    b. **Physical dexterity.** Plasterers and tapers require a steady hand and good hand–eye coordination.

    c. **Math skills.** Plasterers and tapers should have math skills to be able to identify and estimate the quantity of materials needed to complete a job, and accurately estimate how long a job will take to complete and at what cost.

    d. **Physical stamina.** Some workers must stand at their station for extended periods. Continuous standing or activity can be tiring.

    e. **Union/non-union:** Plasterers and stucco masons typically join the Operative Plasterers' and Cement Masons' International Association (OPCMIA). Unions can provide health insurance, job security, pensions, and representation.

9. **Compensation**: In 2019, median pay for plasterers/ stucco masons was $45,440 annually, or $21.85 per hour, not including benefits such as health insurance and retirement. Starting annual pay is closer to $29,360. Top earners earn $78,980 annually or more. Note that generally, union jobs pay higher than non-union jobs, and cities pay higher than rural areas. In 2018, there were 24,180 plasterers/stucco masons in the U.S.

10. **Employment outlook**: Employment of plasterers/ stucco masons is projected to grow 7 percent from 2018 to 2028, about the same as other occupations.

See also: **Trade 3:** Cabinetmaker/woodworker
Trade 7: Drafter
Trade 17: Painter

TRADE 19

Plumber, Pipefitter, and Steamfitter

Indoor plumbing is another wonder of the world. Plumbing includes methods of moving water for irrigation, washing, cooking, and sanitation, which have enormously improved our quality of life. Plumbers install and repair systems to manage water, sewage, and drainage, whereas pipefitters and steamfitters are specialized individuals who install and maintain certain types of pipes for liquids and gases.

1. **History of plumbing:** Plumbing originated during ancient civilizations (such as the early Greek, Roman, Persian, Indian, and Chinese cities) as they developed public baths and needed to provide potable water and wastewater removal for larger numbers of people. In the U.S., Boston pioneered the first water system in the mid-1600's. The first valve-type flush toilet was invented in 1738 by J.F. Brondel. Alexander Cumming patented the flush toilet in 1775, the beginning of the modern toilet. An early plumber gave Queen Elizabeth I the first flushable toilet, but she was scared to use it because of the sounds of rushing water!

2. **What plumbers, pipefitters, and steamfitters do:** Plumbers, pipefitters, and steamfitters install and repair piping fixtures and systems to manage water, sewage, and drainage. *Plumbers* install and repair water, gas, and other piping systems in homes, businesses, and factories. They install plumbing fixtures, such as bathtubs and toilets, and appliances, such as dishwashers and

water heaters. They clean drains, remove obstructions, and repair or replace broken pipes and fixtures and septic systems. *Pipefitters* and *steamfitters* install and maintain pipes that may carry chemicals, acids, and gases. These pipes are mostly in manufacturing, commercial, and industrial settings. Steamfitters specialize in systems that are designed for the flow of liquids or gases at high pressure.

3. **Work environment:** Plumbers, pipefitters, and steamfitters work in factories, homes, businesses, and other places where there are pipes and related systems. Plumbers are often on call for emergencies, so evening and weekend work is common.

4. **Education needed:** A high school diploma or equivalent is typically required to become a plumber, pipefitter, or steamfitter. Vocational-technical schools offer courses in pipe system design, safety, and tool use. They also offer welding courses that are required by some pipefitter and steamfitter apprenticeship training programs.

5. **Training/Apprenticeship:** Most plumbers, pipefitters, and steamfitters learn their trade through a 4- or 5-year apprenticeship. Apprentices typically receive 2,000 hours of paid on-the-job training, as well as some technical instruction, each year. Technical instruction includes safety, local plumbing codes and regulations, and blueprint reading. Apprentices also study mathematics, applied physics, and chemistry. Apprenticeship programs are sponsored by unions, trade associations, and businesses. Most apprentices enter a program directly, but some start out as helpers or complete pre-apprenticeship training programs in plumbing and other trades.

6. **After apprenticeship:** Plumbers, pipefitters, and steamfitters complete an apprenticeship program and pass the required licensing exam to become journey-level workers. Journey-level plumbers, pipefitters, and steamfitters are qualified to perform tasks independently. Plumbers with several years of plumbing experience who pass another exam earn master status.

7. **License/certification**: Most states and some localities require plumbers to be licensed. Although licensing requirements vary, states and localities often require workers to have 2 to 5 years of experience and to pass an exam that shows their knowledge of the trade before allowing plumbers to work independently. Plumbers may also obtain optional certification, such as in plumbing design, to broaden career opportunities. In addition, most employers require plumbers to have a driver's license. Some states require pipefitters and steamfitters to be licensed; they may also require a special license to work on gas lines. Licensing typically requires an exam or work experience or both. Contact your state's licensing board for more information.

8. **Competencies needed**: The following are required competencies for plumbers, pipefitters, and steamfitters.

   a. **Communication skills.** Plumbers, pipefitters, and steamfitters must be able to direct workers, bid on jobs, and plan work schedules. Plumbers also talk to customers regularly.

   b. **Dexterity.** Plumbers, pipefitters, and steamfitters must be able to maneuver parts and tools precisely, often in tight spaces.

   c. **Mechanical skills.** Plumbers, pipefitters, and steamfitters choose from a variety of tools to assemble, maintain, and repair pipe systems.

   d. **Physical strength.** Plumbers, pipefitters, and steamfitters must be able to lift and move heavy tools and materials.

   e. **Troubleshooting skills.** Plumbers, pipefitters, and steamfitters find, diagnose, and repair problems. They also help with setting up and testing new plumbing and piping systems.

9. **Union/non-union:** Plumbers, pipefitters, and steamfitters typically join the United Association of Journeymen and Apprentices of the Plumbing and Pipe Fitting Industry of the

United States, Canada and Australia, often abbreviated as the UA. Unions can provide health insurance, job security, pensions, and representation.

10. **Compensation**: In 2019, median pay for plumbers, pipefitters, and steamfitters was $55,160 annually, or $26.52 per hour, not including benefits such as health insurance and retirement. Starting annual pay is closer to $32,690. Top earners earn $97,170 annually or more. Note that generally, union jobs pay higher than non-union jobs, and cities pay higher than rural areas. In 2018, there were 500,300 plumbers, pipefitters, and steamfitters in the U.S.

11. **Employment outlook**: Employment of plumbers is projected to grow 14 percent from 2018 to 2028, much faster than the average for all occupations. After completing an apprenticeship and becoming licensed at the journey level, plumbers may advance to become a master plumber, supervisor, or project manager. Some plumbers choose to start their own business as an independent contractor, which may require additional licensing.

See also: **Trade 1:** Boilermaker
Trade 11: Heating, Ventilation, Air-conditioning, and Refrigeration Technician
Trade 15: Machinist/Tool and Die Maker
Trade 25: Construction Manager/Project Manager

TRADE 20

Roofer

Buildings are not very useful if they don't have solid roofing! Roofers make sure buildings are safe from rain and sun.

1. **History of roofing:** The evolution of roofing design can be traced far back as 3000 B.C., when the Chinese used clay roof tiles. Roman and Greek civilizations utilized slate and tile in the first century. By the eighth century, thatched roofs became the common form of most areas of Western Europe, and wooden shingles were prevalent in the eleventh century. The asphalt shingle was created around the turn of the 20th century and continues to be the top roofing material for houses.

2. **What roofers do:** Roofers replace, repair, and install the roofs of buildings. Roofers work on new installations, renovations, and roof repair. If needed, roofers also replace old materials with new, solid structures. Roofers deal in many different materials including metal, rubber, polymer, asbestos and tile.

3. **Work environment:** Roofing work can be physically demanding because it involves heavy lifting, as well as climbing, bending, and kneeling, frequently in very hot weather. Roofers may work overtime in order to finish a particular job, especially during busier summer months. Overtime may pay more for union members than for non-union members.

4. **Education needed:** There are no specific education requirements for roofers.

5. **Training needed:** Most on-the-job training programs consist of instruction in which experienced workers teach new workers how to use roofing tools, equipment, machines, and materials. Trainees begin with tasks such as carrying equipment and material and erecting scaffolds and hoists. Within 2 or 3 months, they are taught to measure, cut, and fit roofing materials. Later they are shown how to lay asphalt or fiberglass shingles. Because some roofing materials, such as solar tiles, are used infrequently, it can take several years to gain experience on all types of roofing. As training progresses, new workers are able to learn more complex roofing techniques.

6. **Apprenticeships:** Apprenticeships combine on-the-job training with classroom instruction. A few groups—including the United Union of Roofers, Waterproofers and Allied Workers and some contractor associations—sponsor apprenticeship programs for roofers.

7. **License/certification:** None required.

8. **Competencies needed:** The following are required competencies for roofers.

   a. **Balance.** Roofers should have excellent balance to avoid falling, because the work is often done on steep slopes at significant heights.

   b. **Manual dexterity.** Roofers need to be precise when installing roofing materials and handling roofing tools, in order to prevent damage to the roof and building.

   c. **Physical stamina.** Roofers must have the endurance to perform strenuous duties throughout the day. They may spend hours on their feet, bending and stooping—often in hot temperatures.

   d. **Physical strength.** Roofers often lift and carry heavy materials. Some roofers, for example, must carry bundles of shingles that weigh 60 pounds or more.

e. **Unafraid of heights.** Roofers must not fear working far above the ground, because the work is often done at significant heights.

9. **Union/non-union:** Roofers typically join the United Union of Roofers, Waterproofers, and Allied Workers. Unions can provide health insurance, job security, pensions, and representation.

10. **Compensation:** In 2019, median pay for roofers was $42,100 annually, or $20.24 per hour, not including benefits such as health insurance and retirement. Starting annual pay is closer to $26,540. Top earners earn $70,920 annually or more. Note that generally, union jobs pay higher than non-union jobs, and cities pay higher than rural areas. In 2018, there were 160,600 roofers in the U.S.

11. **Employment outlook:** Employment of roofers is projected to grow 12 percent from 2018 to 2028, faster than the average for all occupations.

See also: Trade 11: Heating, Ventilation, Air-conditioning, and Refrigeration Technician
Trade 13: Ironworker
Trade 16: Millwright

TRADE 21

Sheet Metal Worker

What is sheet metal used for? A better question is: what isn't it used for? It's found in mailboxes, air-conditioning ducts, guardrails, airplanes, cars, fan blades, and food vats. Sheet metal workers make and repair all of these and more!

1. **History of sheet metal workers:** Ancient Egyptian jewelers used materials similar to sheet metal to make jewelry thousands of years ago. The metals suited for rolling into sheets were lead, copper, zinc, iron and later steel. Tin was often used to coat iron and steel sheets to prevent it from rusting. First sheet metal was created manually, then water-powered rolling mills replaced the manual process in the late 17th century. Sheet metals appeared in the U.S. in the 1870s and were used for shingle roofing, stamped ornamental ceilings, and exterior façades. Sheet metal was a highly popular building product until World War II, when metals became needed for the war effort, and the industry struggled. Today, there are about 4,400 fabrication shops in the U.S., and the industry is worth around $20.5 billion.

2. **What sheet metal workers do:** Sheet metal workers fabricate or install products that are made from thin metal sheets.

3. **Work environment:** Sheet metal workers often lift heavy materials and stand for long periods of time. Those who install sheet metal must often bend, climb, and squat. Most work full time.

4. **Education needed:** Sheet metal workers typically need a high

school diploma or equivalent. Those interested in becoming a sheet metal worker should take high school classes in algebra and geometry. Vocational-education courses such as blueprint reading, mechanical drawing, and welding are also helpful. Technical schools may have programs that teach welding and metalworking. These programs help provide the basic welding and sheet metal fabrication knowledge that sheet metal workers need to do their job. Some manufacturers have partnerships with local technical schools to develop training programs specific to their factories.

5. **Training/Apprenticeships:** Most construction sheet metal workers learn their trade through 4- or 5-year apprenticeships, which include both paid on-the-job training and related technical instruction. Apprentices learn construction basics such as blueprint reading, math, building code requirements, and safety and first aid practices. Welding may be included as part of the training. Some workers start out as helpers before entering apprenticeships. Apprenticeship programs are sponsored by unions and businesses. The basic qualifications for entering an apprenticeship program are being 18 years old and having a high school diploma or the equivalent.

6. **After apprenticeship:** After completing an apprenticeship program, sheet metal workers are considered journey workers who are qualified to perform tasks on their own.

7. **License/certification:** Some states require licenses for sheet metal workers. Check with your state for more information. Although not required, sheet metal workers may earn certifications for several tasks that they perform. For example, some sheet metal workers become certified in welding from the American Welding Society (AWS). In addition, the International Certification Board (ICB) offers certification in testing and balancing, HVAC fire life safety, and other related activities for eligible sheet metal workers. The Fabricators & Manufacturers

Association, International (FMA) offers a certification in precision sheet metal work.

8. **Competencies needed**: The following are required competencies for sheet metal workers:

   a. **Detail oriented.** Sheet metal workers must precisely measure and cut, follow detailed directions, and monitor their surroundings for safety risks.

   b. **Dexterity.** Sheet metal workers need good hand–eye coordination and motor control to make precise cuts and bends in metal pieces.

   c. **Math skills.** Sheet metal workers must calculate the proper sizes and angles of fabricated sheet metal to ensure the alignment and fit of ductwork.

   d. **Mechanical skills.** Sheet metal workers use saws, lasers, shears, and presses. They should have good mechanical skills in order to operate and maintain equipment.

   e. **Physical stamina.** Sheet metal workers in factories may spend many hours standing at their workstation.

   f. **Physical strength.** Sheet metal workers must be able to lift and move ductwork that is heavy and cumbersome. Some jobs require workers to push, pull, or lift 50 pounds or more.

9. **Union/non-union:** The union that represents this group is the International Association of Sheet Metal, Air, Rail and Transportation Workers (SMART), which was formed when the Sheet Metal Workers' International Association merged with the United Transportation Union in 2014. Unions can provide health insurance, job security, pensions, and representation.

10. **Compensation**: In 2019, median pay for sheet metal workers was $50,400 annually, or $24.23 per hour, not including benefits such as health insurance and retirement. Starting annual pay is

closer to $29,260. Top earners earn $88,070 annually or more. Note that generally, union jobs pay higher than non-union jobs, and cities pay higher than rural areas. In 2018, there were 143,000 sheet metal workers in the U.S.

11. **Employment outlook**: Employment of sheet metal workers is projected to grow 8 percent from 2018 to 2028, about as fast as for other occupations.

See also: **Trade 11:** Heating, Ventilation, Air-conditioning, and Refrigeration Technician
Trade 13: Ironworker
Trade 16: Millwright

TRADE 22

Surveyor

Surveyors work in the lineage of explorers William Clark and Meriwether Lewis, who traveled across the Western U.S. to Oregon, surveying land and making maps.

1. **History of surveying:** Land surveying is an ancient practice that dates back at least to 1,400 B.C., when the ancient Egyptians used land surveying for the taxation of land plots. Four thousand years ago, Egyptians used measuring ropes, plumb bobs, and other instruments to gauge the dimensions of plots of land. In the Qin dynasty in China, around 200 B.C., the first magnetic compass was created, becoming an essential tool for surveying. In 1086, William the Conqueror ordered a "Domesday Book," which was a record of people who owned land in England and the size of the plots they owned, including each plot's boundaries, property elements, and inhabitants.

2. **What surveyors do:** Surveyors make precise measurements to determine property boundaries.

3. **Work environment:** Surveying involves both fieldwork and office work. When working outside, surveyors may stand for long periods and often walk long distances, sometimes in bad weather. Most work full time.

4. **Education needed:** Surveyors typically need a bachelor's degree because they work with sophisticated technology and math. Some colleges and universities offer bachelor's degree programs

specifically designed to prepare students to become licensed surveyors. Many states require individuals who want to become licensed surveyors to have a bachelor's degree from a school accredited by the Accreditation Board for Engineering and Technology (ABET). A bachelor's degree in a closely related field, such as civil engineering or forestry, is sometimes acceptable as well. An associates degree may be sufficient in some cases with additional training.

5. **Training needed:** Surveying professionals must have strong mathematical skills in order to understand the complexities of calculating averages, measuring angles and computing land mass areas. Surveying requires specialized equipment, such as high-precision and electromechanical instruments and global positioning technologies, to acquire spatial data, perform data reduction, analyze measurements, and make data adjustments. A bachelor's degree in surveying engineering provides theoretical and practical knowledge of geospatial systems, spatial data collecting, global positioning systems and other surveying technologies. Through hands-on assignments students learn to use modern technology to analyze hydraulic systems, land development sites, highways and structures. A bachelor's degree typically takes four years to complete. Course topics usually include advanced math, data collection systems, survey computations, legal aspects of surveying, and boundary surveying principles.

6. **License/certification:** All 50 states and the District of Columbia require surveyors to be licensed before they can certify legal documents that show property lines or determine proper markings on construction projects. In order to become licensed, most states require approximately 4 years of work experience and training under a licensed surveyor after obtaining a bachelor's degree. Other states may allow substituting more years of work experience and supervised training under a licensed surveyor in place of education. The amount of work experience required varies by state. Check with your state for more information.

Although the process of obtaining a license varies by state, the National Council of Examiners for Engineering and Surveying (NCEES) has a generalized process of four steps: Complete the level of education required in your state, pass the Fundamentals of Surveying (FS) exam, gain sufficient work experience under a licensed surveyor, and pass the Principles and Practice of Surveying (PS) exam. Most states also have continuing education requirements for surveyors to maintain their license.

7. **Competencies needed**: The following are required competencies for surveyors.

  a. **Communication skills.** Surveyors must provide clear instructions to team members, clients, and government officials. They also must be able to follow instructions from architects and construction managers, and explain the job's progress to developers, lawyers, financiers, and government authorities.

  b. **Detail oriented.** Surveyors must work with precision and accuracy because they produce legally binding documents.

  c. **Physical stamina.** Surveyors traditionally work outdoors, often in rugged terrain. They must be able to walk long distances and for long periods.

  d. **Problem-solving skills.** Surveyors must figure out discrepancies between documents showing property lines and current conditions on the land. If there were changes in previous years, they must discover the reason behind them and reestablish property lines.

  e. **Time-management skills.** Surveyors must be able to effectively plan their time and their team members' time on the job. This is critical when there are pressing deadlines or while working outside during winter months, when daylight hours are short.

f.   Visualization skills. Surveyors must be able to envision new buildings and altered terrain.

8. **Union/non-union:** Surveyors can join the International Union of Operating Engineers (IUOE) that represents operating engineers, who work as heavy equipment operators, mechanics, and surveyors in the construction industry. Unions can provide health insurance, job security, pensions, and representation.

9. **Compensation:** In 2019, median pay for surveyors was $63,420 annually, or $30.49 per hour, not including benefits such as health insurance and retirement. Starting annual pay is closer to $36,110. Top earners earn $104,850 annually or more. Note that generally, union jobs pay higher than non-union jobs, and cities pay higher than rural areas. In 2018, there were 49,200 surveyors in the U.S.

10. **Employment outlook:** Employment of millwrights is projected to grow 5 percent from 2018 to 2028, as fast as average for all occupations.

See also: Trade 7: Drafter
Trade 12: Heavy Equipment Operator
Trade 15: Machinist/Tool and Die Maker

TRADE 23

Telecommunication Technician

Setting up systems that help people of all generations communicate; telecommunication technicians operate from a long history of technologies: telegraph, telephone, radio, television, and — today — wireless technologies.

1. **History of telecommunication technicians:** The history of telecommunication began with the use of smoke signals and drums in Africa, Asia, and the Americas. In the 1790s, the first fixed semaphore systems emerged in Europe. In the 1800s, electrical telecommunication systems started to appear, beginning with the telegraph and developing into telephone, radio, and television. Today, telecommunications technicians work on satellite communication, digital telephones, the Internet, and wireless technologies.

2. **What telecommunications technicians do:** Telecommunications equipment installers and repairers set up and maintain devices that carry communications signals. Telecommunications technicians typically establish communications systems by installing, operating, and maintaining voice and data telecommunications network circuits and equipment; plan network installations by studying customer orders, plans, manuals, and technical specifications; establish voice and data networks by running, pulling, terminating, and splicing cables; install telecommunications

equipment, routers, switches, multiplexors, cable trays, and alarm and fire-suppression systems; verify service by testing circuits, equipment, and alarms; and maintain networks by troubleshooting and repairing outages, testing network back-up procedures, and updating documentation.

3. **Work environment:** Telecommunications equipment installers and repairers generally work in central offices or electronic service centers. They also work in the homes and offices of customers. Some technicians travel frequently to installation and repair sites.

4. **Education needed:** Telecommunications technicians typically need a postsecondary education in electronics, telecommunications, or computer networking. Generally, postsecondary programs include classes such as data transmission systems, data communication, AC/DC electrical circuits, and computer programming. Most programs lead to a certificate or an associate's degree in telecommunications or related subjects.

5. **Training needed:** Once hired, telecommunications technicians receive on-the-job training, typically lasting a few weeks to a few months. Training involves a combination of classroom instruction and hands-on work with an experienced technician. In these settings, workers learn the equipment's internal parts and the tools needed for repair. Technicians who have completed postsecondary education often require less on-the-job instruction than those who have not. Some companies may send new employees to training sessions to learn about equipment, procedures, and technologies offered by equipment manufacturers or industry organizations. Because technology in this field constantly changes, telecom technicians must continue learning about new equipment over the course of their careers.

6. **License/certification:** None required.

7. **Competencies needed:** The following are required competencies for telecommunication technicians.

a. **Color vision.** Telecom technicians work with color-coded wires, and they need to be able to tell them apart.

b. **Customer-service skills.** Telecom technicians who work in customers' homes and offices should be friendly and polite. They must be able to teach people how to maintain and operate communications equipment.

c. **Dexterity.** Telecom technicians' tasks, such as repairing small devices, connecting components, and using hand tools, require a steady hand and good hand–eye coordination.

d. **Mechanical skills.** Telecom technicians must be familiar with the devices they install and repair, with their internal parts, and with the appropriate tools needed to use, install, or fix them. They must also be able to understand manufacturers' instructions when installing or repairing equipment.

e. **Troubleshooting skills.** Telecom technicians must be able to troubleshoot and devise solutions to problems that are not immediately apparent.

8. **Union/non-union:** Telecommunication Technicians frequently belong to the union Communications Workers of America (CWA). Unions can provide health insurance, job security, pensions, and representation.

9. **Compensation:** In 2019, median pay for telecom technicians was $57,910 annually, or $27.84 per hour, not including benefits such as health insurance and retirement. Starting annual pay is closer to $33,090. Top earners earn $85,620 annually or more. Note that generally, union jobs pay higher than non-union jobs, and cities pay higher than rural areas. In 2018, there were 232,900 telecom technicians in the U.S.

10. **Employment outlook:** Employment of telecom technicians is projected to decrease 6 percent from 2018 to 2028.

See also: **Trade 7:** Drafter
Trade 8: Electrician
Trade 22: Surveyor

TRADE 24

Tractor Trailer Truck Drivers

Since the 1970s, truckers have been viewed like a modern version of cowboys—lots of work to do, long periods of time alone, and a dangerous occupation. If hitting the road for long periods of time bringing freight across the country sounds good to you, read on!

1. **History of tractor trailer truck driving:** There has always been a need for hauling freight and supplies. Freight was transported by horse and wagon, by canals, and then by railroads. The invention of the automobile prompted an increase in road quality and enabled long-distance driving. In 1914, a blacksmith in Detroit built a detachable trailer for a Ford automobile and called the new invention a "semi-trailer." Truck technology continued to improve, as did the quality of roads, and the specialization of tractor trailer truck drivers.

2. **What tractor trailer truck drivers do:** Tractor-trailer truck drivers drive long distances. They are also responsible for securing cargo, inspecting their trailers and reporting problems, reporting road incidents, logging all working hours, and keeping their trucks and associated equipment clean and in good working order. Certain cargo requires drivers to adhere to additional safety regulations. Some heavy truck drivers who transport hazardous materials, such as certain chemicals, must take special precautions when driving and may carry specialized

safety equipment in case of an accident. Some truck drivers are owner-operators who buy or lease trucks and go into business for themselves.

3. **Work environment:** Working as a long-haul truck driver can be a lifestyle choice because drivers can be away from home for days or weeks at a time, often alone. The job is physically demanding as well, both from driving or from loading/unloading cargo. Drivers work long hours, and also have limits on driving: drivers may not work more than 14 hours straight, including up to 11 hours driving and the remaining time doing other work, such as unloading cargo. Between working periods, drivers must have at least 10 hours off duty. Drivers often work nights, weekends, and holidays.

4. **Education needed**: Most companies require truck drivers to have a high school diploma or equivalent. Many prospective drivers attend professional truck driving schools, where they take training courses to learn how to maneuver large vehicles on highways or through crowded streets. During these classes, drivers also learn federal laws and regulations governing interstate truck driving. These programs are offered privately and through community colleges. Typically, they last 3-6 months.

5. **Training needed:** After completing truck-driving school and being hired by a company, drivers normally receive several weeks of on-the-job training. During this time, they drive a truck accompanied by an experienced mentor-driver in the passenger seat.

6. **License/certification**: All long-haul truck drivers must have a commercial driver's license (CDL). Qualifications for obtaining a CDL vary by state but generally include passing a knowledge test and a driving test. Drivers can get endorsements to their CDL that show their ability to drive a specialized type of vehicle. Truck drivers transporting hazardous materials (hazmat) must have a hazardous materials endorsement, which requires passing

an additional knowledge test and a background check. Federal regulations require CDL drivers to maintain a clean driving record and pass a physical exam every two years. They are also subject to random testing for drug or alcohol abuse.

7. **Competencies needed**: The following are required competencies for tractor-trailer truck drivers.

8. **Hand-eye coordination**. Drivers of heavy trucks and tractor-trailers must be able to coordinate their legs, hands, and eyes simultaneously so that they will react appropriately to the situation around them and drive the vehicle safely.

9. **Hearing ability**. Truck drivers need good hearing. Federal regulations require that a driver be able to hear a forced whisper in one ear at 5 feet away (with or without the use of a hearing aid).

10. **Physical health**. Federal regulations do not allow people to become truck drivers if they have a medical condition, such as high blood pressure or epilepsy, which may interfere with their ability to operate a truck.

11. **Visual ability**. Truck drivers must be able to pass vision tests. Federal regulations require a driver to have at least 20/40 vision with a 70-degree field of vision in each eye and the ability to distinguish the colors on a traffic light.

12. **Union/non-union**: The Freight Division of the International Brotherhood of Teamsters represents truckers. Most trucking jobs, however, are non-union. Unions can provide health insurance, job security, pensions, and representation.

13. **Compensation**: In 2019, median pay for tractor trailer truck drivers was $45,260 annually, or $21.76 per hour, not including benefits such as health insurance and retirement. Starting annual pay is closer to $29,130. Top earners earn $66,840 annually or more. Note that generally, union jobs pay higher than

non-union jobs, and cities pay higher than rural areas. In 2018, there were 1,958,800 tractor-trailer truck drivers in the U.S.

14. **Employment outlook**: Employment of tractor-trailer truck drivers is projected to grow 5 percent from 2018 to 2028, average for all occupations.

See also: **Trade 6:** Diesel Technician
Trade 12: Heavy Equipment Operator
Trade 15: Machinist/Tool and Die Maker

TRADE 25

Construction Manager/Project Manager

Once you've learned your trade, you may be interested in overseeing the work of others. Typically, construction managers are called foremen or forewomen. Construction managers are responsible for organizing the overall project and supervising and training employees in their charge.

1. **History of construction managers:** Prior to World War 2, general contractors typically ran construction projects. In the 1950s and 1960s, as postwar construction projects boomed, cost and time overrun became a major issue. The specialty of construction manager was created to be the point person to ensure all work was done within cost and time limitations. Training programs then emerged for construction managers.

2. **What construction managers do:** Construction managers are experts in certain trades who oversee the management of that work on the jobsite. They are responsible for training employees, providing appropriate documentation and instruction to workers, ensuring safe and appropriate use of equipment to employees, maintaining the employee schedule, managing costs, managing quality, making decisions, and communicating progress on the project. Some construction managers use cost-estimating and

planning software to determine costs and the materials and time required to complete projects.

3. **Work environment:** Many construction managers have a main office but spend most of their time working out of a field office at a construction site, where they monitor the project and make daily decisions about construction activities. The need to meet deadlines and respond to emergencies often requires them to work many hours.

4. **Education needed:** Typically, construction managers are construction workers with many years of experience in a particular trade. They generally need to have a bachelor's degree and learn management techniques through on-the-job training. Large construction firms increasingly prefer candidates with both construction experience and a bachelor's degree in a construction-related program in construction science, construction management, architecture, or engineering. Bachelor's degree programs typically include courses in project control and management, design, construction methods and materials, cost estimation, building codes and standards, contract administration, mathematics, and statistics.

5. **Training needed:** In addition to previous work in the construction industry, there are various internships, cooperative education programs, and other opportunities to train with experienced construction managers. Some construction managers become qualified solely through extensive construction experience, spending many years in carpentry, masonry, or other construction specialties.

6. **License/certification:** Some states require licensure for construction managers. Many local colleges and community colleges provide certification. The Construction Management Association of America (CMAA) awards a national Certified Construction Manager (CCM) designation to workers with required experience and a passing exam score.

7. The American Institute of Constructors awards the Associate Constructor (AC) and Certified Professional Constructor (CPC) designations to candidates who meet its requirements and pass their exams.

8. **Competencies needed**: The following are required competencies for construction managers.

   a. **Leadership skills**. Construction managers must effectively delegate tasks to construction workers, subcontractors, and other lower level managers.

   b. **Technical skills**. Construction managers must know their trade. They need to know construction methods and technologies and must be able to interpret contracts and technical drawings.

   c. **Analytical skills**. Construction managers plan project strategies, handle unexpected issues and delays, and solve problems that arise over the course of the project.

   d. **Business skills**. Construction managers address budget matters and coordinate and supervise workers. Choosing competent staff and establishing good working relationships with them is critical.

   e. **Customer-service skills**. Construction managers are in constant contact with owners, inspectors, and the public. They must create good working relationships with these people and ensure their needs are met.

   f. **Decision-making skills**. Construction managers choose personnel and subcontractors for specific tasks and jobs. Often, these choices must be made quickly to meet deadlines and budgets.

   g. **Initiative**. Self-employed construction managers generate their own business opportunities and must be proactive in finding new clients. They often market their services and bid on jobs, and they must also learn to perform special home-improvement projects, such as installing mosaic glass tiles, sanding wood floors, and insulating homes.

h. **Speaking skills**. Construction managers must give clear orders, explain complex information to construction workers and clients, and discuss technical details with other building specialists, such as architects. Self-employed construction managers must get their own projects, so the need to sell their services to potential clients is critical.

i. **Time-management skills**. Construction managers must make sure the work meets deadlines so that the next phase of work can continue.

j. **Writing skills**. Construction managers must write proposals, plans, and budgets, as well as document the progress of the work for clients and others involved in the building process.

9. **Union/non-union:** As a manager, construction managers do not have a separate union; however, they can still be part of their own trade union (such as for carpenters or electricians).

10. **Compensation**: In 2019, median pay for construction supervisors was $95,260 annually, or $45.80 per hour, not including benefits such as health insurance and retirement. Starting annual pay is closer to $56,140. Top earners earn $164,790 annually or more. Note that generally, union jobs pay higher than non-union jobs, and cities pay higher than rural areas. In 2018, there were 471,800 construction managers in the U.S.

11. **Employment outlook**: Employment of construction managers is projected to grow 10 percent from 2018 to 2028, faster than the average for all occupations. Construction managers are expected to be needed to oversee the anticipated increase in construction activity over the coming decade.

See also: **Trade 2:** Bricklayer/Mason
Trade 4: Carpenter
Trade 8: Electrician
Trade 19: Plumber, Pipefitter, and Steamfitter

If you're interested in this construction trade, check out *Millennials' Guide to Management & Leadership* by Jennifer P. Wisdom. It's another Millennials' Guide that might have just what you need.

SECTION II

Skills and Abilities for Being Successful in Construction Trades

SKILL/ABILITY 1

Physical Ability

All construction trades require some physical ability because of the physical size and weight of materials and the manual dexterity necessary to lift, move, or hold materials. You don't have to be a big person, but it's important to consider a match between your physical abilities and what the job requires.

1. **Good health and strength.** This is often taken for granted. Millennials considering the construction trades needs to be in good physical condition. Smoking, excessive drinking and drug use shorten careers and lives dramatically. The human body can only hold up to so much physical abuse.

2. **Underlying physical conditions.** Some Millennials have allergies, or asthma, or other physical conditions that may make some trades challenging. Take these into consideration when choosing a trade. Carpenters, for example, work in dusty environments. A person with allergies may find this trade to be difficult or impossible to work in. However, that same person may do well as an electrician in most conditions.

3. **Being of a physical size to be able to do manual labor.** While many construction tradespeople are bigger in stature, it is not a requirement of the job. Having said that, keep in mind that some trades, such as laborer or ironworker, require physical strength on a constant basis. In those types of occupations, it helps to be a big, strong person. Brute force is always in demand

at a construction job simply because of the weight of the materials being used in building. There is always a place for physical strength.

4. **Being nimble and detail oriented.** In contrast to the manual labor jobs mentioned above, in many other trades, such as finish carpentry, electrician, or surveyor, being nimble or small may actually work to your advantage to squeeze into tight places and to work with small materials.

5. **Being able to utilize one's personal strengths.** For Millennials entering the trades, look to yourself for the strength. Ask yourself what are you good at? Look for a trade that suits your basic talents and natural abilities. Are you capable of working all day in a harsh environment? Can you consistently work all day long?

6. **Safety.** Safety is a constant concern in construction. Construction workers are often seriously injured as a result of an accident. Sadly, sometimes workers are killed. Keep in mind that injuries can occur without being involved in an accident: Repetitive motion injuries, sprains, cuts, bruises, and pulled muscles are all common everyday occurrences. In addition, it is easy to pull out one's back doing even some of the simplest movements. You must take care to guard your health!

See also: Skill/Ability 2: Mechanical ability
Skill/Ability 3: An aptitude for hand skills with tools
Skill/Ability 10: Dealing with unfavorable conditions

Take Action: Is physical ability a primary strength of yours? If so, check out the jobs for which this is a key component, including **Trade 13:** Ironworker and **Trade 14:** Laborer.

SKILL/ABILITY 2

Mechanical Ability

Working in the trades is without a doubt a practical, hands-on career path. There is theory as to how things work and the way things were planned and designed, but ultimately it comes down to actual hands-on labor that makes construction happen. There is what was planned and then there is making it happen. The construction trades are heavily involved on the making it happen side of the projects. Starting off everyone in a trade needs to learn the mechanical skills and fundamental skills of the trade.

1. **What is mechanical ability?** Mechanical ability is simply the ability to skillfully use tools or machines. Some Millennials have more mechanical ability than others. Some Millennials are "book smart" and work in the world of theories and ideas. In contrast, construction tradespeople of all generations need to understand both the ideas to plan out the work and also be able to do hands-on work to install, create, and repair things.

2. **Mechanical ability includes understanding how things work and move independently and in relation to each other.** For example, understanding how weights and pulleys, levers, acceleration, and gears work.

3. **Are you curious about how things work in your home or how machinery works?** If you are the kind of person who likes to think through how systems work as opposed to just some magic behind the scenes, then you are probably a good candidate for a career in the construction trades.

4. **Mechanical ability is an area where a smaller person can outperform a bigger stronger person.** The good thing is that if one has the basic mechanical aptitude, you can learn a lot to improve your skills. This will make a worker faster and more productive while at the same time not having to work as hard to produce the same results.

5. **It's also important to pay attention to what you enjoy.** If assembling furniture seems fun, you probably have a good basic mechanical ability to start off with. Some Millennials simply dread the thought of assembling anything. Others enjoy putting things together, and can often complete most of the assembly without instructions

See also: Skill/Ability 3: An aptitude for hand skills with tools
Skill/Ability 12: Determination

Take Action: Find a mechanical aptitude test online and see how you score. Or find a video on test preparation for a mechanical aptitude test, and see if you're interested in the concepts, whether you can answer the questions, and whether you enjoy trying to answer them. If you find yourself interested or intrigued by mechanics and technology, construction might be a good industry for you!

SKILL/ABILITY 3

An Aptitude for Hand Skills with Tools

What separates tradespeople from other workers is skill. Skill is knowledge plus practice. First you must learn and know what to do, and then you must practice the knowledge and manual dexterity until you become good at it. Very skilled professionals like musicians practice the same activities over and over until they become excellent. You can do the same thing with construction trades, whether it's measuring, cutting, hammering, joining, or other skills. Know what to do, and then practice doing it until you are excellent.

1. **Build from learning the simplest tasks first.** Often Millennials want to do the advanced level skills first, but in any trade or field it is always best to learn the basics first. You have to crawl before you can walk.

2. **Learning the fundamentals of a trade is essential.** Although they may seem simple and rudimentary, it is the thorough knowledge of the basics that allow for one to learn the more advanced skills.

3. **Skills create opportunities.** The more you develop your skills, the more opportunities you'll have for employment and ultimately for long-term success.

4. **Patience is key.** Ask yourself if you are patient to learn a new skill required to do a home-improvement project? Are you willing to

put in the required effort to learn the skill and see the project through to the end? It's great if you can learn skills quickly, but everyone encounters a skill they have to really practice, and in these situations, patience will get you through it.

5. **Curiosity will help you learn.** The more you are curious about why things didn't go right, or what you did this time that made it work, the more you will learn. And the more curiosity you have about how a large structure gets built, or how a big machine gets made, the more you will expand your skills into new areas.

6. **As you develop skills, you learn more about the industry in general.** As you watch other tradespeople practice skills, you also learn about their values, how they handle mistakes, how they treat other people, and what it takes to be a successful professional in the field. These are often skills you can't learn in a classroom!

See also: Skill/Ability 2: Mechanical ability
Skill/Ability 7: Logical thinking and planning
Success Tip 3: Build on your Strengths

Take Action: What kind of hobbies do you enjoy? Many Millennials with skills for creating art or carving, for example, could also feel comfortable using their hands in construction work.

SKILL/ABILITY 4

Ability to Understand Drawings and Diagrams

Have you ever seen a construction plan (also called a blueprint)? Construction plans are architectural drawings that explain the details of a project, including dimensions, parts, placement, and materials for each project to assure the project is completed correctly. The symbols and lines represent walls, landscape features, wiring, and plumbing. Being able to truly understand drawings and diagrams, however, takes the ability to visualize what the drawings are representing and apply the information from the drawing to getting the work done.

1. **How good are you at visualizing?** Do you enjoy reading or looking at diagrams and imagining how what they represent in real life? If so, that is a good sign that this is a strong ability of yours.

2. **Formal training can increase your skills** if this isn't a strength for you. As with many things in life, practice makes perfect.

3. **The more experience you have in seeing construction plans** and then watching the structure take shape, the easier you will be able to visualize everything yourself.

4. **Millennials have a distinct advantage** in the area of construction plan reading. Construction plans are no longer produced

by hand and are all produced using software, such as Visual Architect, InDesign, Archicad, or other building information modeling software. These systems are more similar to communications Millennials are comfortable with (like cellphones and video games) instead of the old-fashioned pencil and paper designs. Once you understand the virtual world of the video game, it is the same as understanding the virtual world being presented in the new three-dimensional drawings.

5. **Drawings and diagrams are usually presented in a formal written language,** whereas workers on jobsites often use slang or trade names. You might have to learn both languages at the same time!

6. **Hours put in after the workday has finished to learn about blueprints pay huge dividends later on.** The learning process is hard because there is so much information and so many things happening at once on a construction project. But the effort is always worth it.

7. **Take classes about reading construction plans.** Many are available online. Learning to read construction plans is very important. Millennials who want to rise up to the positions of higher responsibility must be skilled at reading construction plans!

See also: Skill/Ability 5: Simple math
Skill/Ability 7: Logical thinking and planning

Take Action: Do an Internet search on Building Informational Modelling (BIM). Understanding how this technology is being applied will give you an advantage as someone starting out in the construction trades. The older tradespeople will value the input from a younger person who is capable with the latest technology. This is especially true if the younger person is able to explain and coach the older, more experienced worker on the new technological systems. It is a big advantage if you are able to teach someone in simple terms about newer electronic systems and how to navigate them.

SKILL/ABILITY 5

Simple Math

Although simple math is often taken for granted, it is one of the most important skills for any tradesperson. Mechanical trades such as heating, ventilation, air-conditioning and refrigeration technicians (HVACR); electrician; plumbing; and pipe fitting all involve mathematical calculations. For carpenters, math is an everyday occurrence. In short, tradespeople use simple math constantly. It's in your best interests to get comfortable with math!

1. **Simple math and its importance to all construction trades cannot be over emphasized.** Every construction trade requires the use of math—if not at the basic level, then certainly at the higher levels or at the supervisory levels. Some trades, such as electricians and carpenters, will use more math than others, but all trades rely on simple math to accomplish the work they do. Try to remember that all of construction is using mathematics in an applied situation. The trades are looking to solve and issue, not merely finding an answer for a test question. The questions are real and so the math is real.

2. **Keep in mind that the math required in the construction trades is not the same as the math required for engineering calculations.** Most often, the math required is that of high school level or less! However, it is applied math. The calculations and arithmetic are being done for a specific reason. The calculations must be correct. This math is not being done to pass a test, it is being done to help move a project forward.

3. **Understand that math is not just for "smart" people or people who are "good" at math.** Construction often doesn't require more than high school level math. It's different from high school, though, in that it's applied to real situations on the jobsite. This applied difference makes it more interesting!

4. **The hardest part of the math work in construction trades is not the actual math, but it is understanding how the equations and processes apply to the work of the trade.** The actual math being used is not difficult, but it needs to be done carefully.

5. **In school, apprenticeship, and on the jobsite, observe when math skills are applied.** If you see people making calculations, ask questions so you can learn.

6. **Math is not just useful for construction** – it's also important for determining wages and the costs of day-to-day living. You will never regret being good at math! Everyday life continuously presents math problems to be solved. Just think about the percentage for interest being charged on a car loan, or the overall cost of insurance if paid out as a monthly bill.

7. **You may want to also consider your confidence in your math skills.** Many Millennials never developed confidence in their math skills because they only learned skills to take tests. Once you apply math to everyday life, you will develop more confidence in your math skills.

8. **For tradespeople, math affects the work they do.** Not all of the workers are doing the math, but those that are organizing and supervising various aspects of the project understand the math and how to apply it. So if you hope to advance from the beginning levels of a trade to the higher level of a foreman/forewoman or supervisor, math becomes an essential skill.

See also: **Skill/Ability 10:** Ability to overcome fears
Skill/Ability 13: Determination

Take Action: Dedicate some time to observing how math is ever-present in our world, especially construction trades. Commit to becoming better at math. Review fundamental math skills, including with online basics such as www.math.com. If you did not learn or understand math in school, it is not too late to learn or re-learn math so that you can improve your math skills. Just like any other skill, math abilities will improve with practice. Don't make excuses . . . make progress!

SKILL/ABILITY 6

Good Reading Skills

Reading is one of the most essential skills of life! Everyone who can read can always find information to help out any situation. The enormous strength of the internet is that you can learn instantly about any subject. Of course, YouTube is a tremendous help too, but not everything is on a video!

1. **Reading can help you learn the vocabulary of construction.** As with any other topic, it is important to learn the "language" of the field. How are things expressed and what is being communicated? If someone said, "I am nailing the studs to the header," it might sound odd. But what they actually mean is that they are going to nail the short framing studs above a door or window. All of the trades have their own separate languages that have evolved over time. Learn yours.

2. **It's important to be able to read at a high school level or higher.** Most of construction reading is at the high school level—not at a level of sophisticated literature. Most of the time the engineers and architects are trying to convey ideas in simple terms.

3. **You have to be able to read to follow written directions.** It is important to read and clearly understand new mechanical systems as defined by the manufacturer. Failure to follow the instructions for a product and its application often results in a void of the warranty from the manufacturer and could result in injury or even death.

4. **Certainly, if someone wants to move up in their career, then reading will play an important role.** Accepting responsibilities comes with lots of learning. A vast amount of time can be saved by reading up on many topics, such as management, leadership, estimating, construction cost management, and so on.

5. **If your reading skills are not all that good, or you struggle a bit with reading, that is okay.** The thing to do is to practice. Just like any other skill, reading can be improved with practice. Start with simple things like reading the sports pages or the news. Read simple short stories and books about topics that interest you. Advance onto subjects more closely aligned with your chosen field.

6. **Some adult learning centers also offer reading classes.** Just as a point, there is no need to be embarrassed. If you are having difficulty with something and you seek help or try to improve, then this is something to be proud of. It is one thing to know you need help with something, it is another thing to decide that you will not try to fix an issue that you are having.

> **See also: Skill/Ability 4:** Ability to understand drawings and diagrams
> **Skill/Ability 9:** Ability to overcome fears

Take Action: Develop a habit of looking up words you don't know in an online dictionary. As you clarify the meanings and how the words are applied, you will improve your reading skills.

SKILL/ABILITY 7

Logical Thinking and Planning

The highest paid skill of all is the ability to think. Being able to think through a process is an essential skill for Millennials looking to advance in any career, including construction. Over the years, there have been many tradesmen and tradeswomen who have done well for their employers by being able to think through a situation and solve problems throughout a jobsite.

Is this a strength of yours?

1. **Some Millennials view those who work in construction as less smart than people who work in other fields or in desk jobs.** Hands-on construction workers, however, have applied their experience and logical thought processes to make tremendous strides for production and cost savings as they apply the skills and knowledge of their trade. This is always a "win-win" for everyone involved.

2. **A big part of planning is to have some experience to help predict what is most likely to happen as a project goes forward.** Even from the very basic levels experience will help to guide the sequence of events in planning out a project. Often people will say: "If I only knew such and such, then I would have done this differently." The experience factor is, that in the trades, the situations often repeat, and so the tradespeople are able to fall back

on the experiences they have had to know what went well, and what could be done differently. This experience helps them to think through the work process and also with the logical planning of a project.

3. **A good way to improve your skills in logical thinking and planning is to observe others in action.** Learn their approach to the challenges of the job they are tasked with. Try to determine what resources they use to engage and ultimately solve the challenges of their work.

4. **Mentors can accelerate your learning curve.** If you can, ask the more experienced people to mentor you as you learn your trade.

5. **Think strategically!** Put yourself in the position of the person who has to accomplish a given task. Think how is it going to work for that person? Asking yourself: "What could be done to make this easier" often yields very effective results.

6. **Often supervisors will not care how difficult a task is to accomplish.** This is a mistake. The easier it is to accomplish a task, the more cost effective it is. So efficiency pays at the bottom line.

7. **Always allow for others' input,** as they will often see something or come up with a helpful idea which you not have thought of yet.

8. **If you hope to go into business for yourself,** thinking things through and planning are essential skills to get things started. Constant improvement will be required if the business is to grow and thrive.

See also: **Skill/Ability 8:** Organizational skills
Skill/Ability 12: Determination

Take Action: The best way to learn logical thinking and planning as applied to construction is to dedicate time and effort into studying how things are done. Read up on how the current processes came to be and how some methods are being improved by technology.

SKILL/ABILITY 8

Organizational Skills

Trades need to work efficiently. Every part of a project costs money. Often the costs of a construction project range into the millions and sometimes billions of dollars. At every level of every trade, staying organized will help for efficiency and will help to keep costs down.

1. **One of the key advantages to organization in a workspace is often overlooked.** Safety is always a concern wherever working with tools. Being organized and maintaining good housekeeping practices will always help to promote safety. From an individual workspace right on up to a large-scale project, organization and good housekeeping always pay off.

2. **Good quality tradespeople keep their tools in order as a matter of pride.** It is almost like a brand. When they set about a task, it looks as if it were easy to do, because everything is ready to go. Having to clean or repair tools prior to use delays a project instantly. Sometimes materials need to be protected from damage. Keeping the end result in mind is important to the success of the project. Keeping things in good order allows for work to be done without hesitation. This helps to get the job done quickly.

3. **Keeping materials organized:** Very often this is the job of the lead workers or supervisors. Materials must be kept in order and out of the way of progress, and they must also be easily accessible when needed. Construction workers often wear large tool belts. The reason for this is so that they do not have to go back to their

tool box to get tools that they know they will use often. The tool belt saves time and effort and allows the worker to perform a task effectively.

4. **Part of being skillful at a trade is doing things with little or no difficulty.** Laying out the work process and having the work area ready so that the tasks can be accomplished in an organized manner is important.

5. **There is a lot to be said for a clean workspace.** It is very hard to produce in a cluttered or dirty environment. This is especially true as the projects near the finishing stages. Even on a rough site, there is an organization of materials and labor happening. Without organization, it would be impossible to conduct the project with good results.

See also: **Skill/Ability 4:** Ability to Understand Drawings and Diagrams
Skill/Ability 7: Logical thinking and Planning
Skill/Ability 12: Determination

Take Action: Take some time to observe how different professions organize and plan activities to be efficient and effective. For example, take look behind the counter at a fast food restaurant or a delicatessen. See how they set up to prepare the food to be served, how each person has specific task, the tools to prepare the food are in order, all the wrappers and packages are nearby, and so on. The same thing must happen in all businesses. The whole process must be thought through. Now look at a trades profession in the same way. How do workers organize their tools, their jobsite, their activities? How do they communicate with each other? What measures do they take to ensure safety? Tradespeople set themselves up for success just like any other profession.

Ability to Overcome Fears

The construction trades take newcomers out of their comfort zone very quickly. As a general rule, the bigger the project, the more danger there is. Working at heights is common. Many of the trades will end up working high off the ground. It's important for you to be able to overcome your fears to be successful. How? Read on…

1. **It is important to know the hazards of each position you might be in,** safety measures you can take in advance, and responses if something goes wrong. These should be addressed on the job site, but they are usually addressed in a general way, not specifically for Millennials who have fears in these areas. It's also a good idea to read up on safety and study outside of work, especially if you have fears in these areas.

2. **It's also important to recognize other fears that may not be related to physical safety.** These can include fears of speaking up, being wrong, or taking chances. It's good to know this about yourself so you can work on overcoming them.

3. **Proper training will help to eliminate some of the fears that newer people encounter.** For example, working in an elevated working platform is very scary and intimidating at first, but with proper training and careful operation the fears will subside and w worker will gain confidence in the machine and its workings.

4. **Safety training is often viewed as boring, but it is actually some of the best education ever provided.** It helps the worker to understand the dangers of the job, and the proper safety measure that should be taken. Don't pass up opportunities to take safety training.

5. **Experience is a great teacher, and a source to gain confidence.** Many fears are overcome once things are done properly with good training and safety practices in place.

6. **You can decide if you want to work in a trade that will challenge you to confront your fears, or if you prefer to generally stay safe from them.** For example, if you are afraid of heights, you can choose to work as a heating, ventilation, air-conditioning and refrigeration technician without too many challenges to your fears, but sometimes you will have to climb a ladder and go up on a roof.

7. **If you have a fear of heights or working near edges:** Falls kill more construction workers than any other hazard, but there is now widespread use of harnesses and fall protection safety devices. It should be considered when entering into the construction field that height is a constant challenge. Very few people have no fear of heights. Some will be okay no matter what. But if you have a fear of heights, then you will need to overcome that fear or control it for many of the situations you will be in. Safety training and practice can help.

8. **If you have a fear of speaking up:** Getting more comfortable talking in small groups and one-on-one can help you increase your confidence talking in front of larger groups. Toastmasters groups provide safe opportunities to practice public speaking and to address your fears.

9. **If you have a fear of taking chances or being wrong:** Remember we only grow by doing. Practice skills outside of work to the extent you can, and ask for feedback from more senior and

experienced colleagues. Find mentors who can walk you through challenging activities, and recognize how far you've already come in taking chances!

10. **If you have a fear of working with potentially dangerous tools**: Power tools have added speed and efficiency to the construction trade, but they must be handled carefully. If a tool can drill a hole into concrete or cut through steel, then obviously it could hurt the operator or others nearby. As a precaution, no one should ever operate a power tools or any other tool without training and safety instruction. Unfortunately, safety training and safe tool operation are too often taken for granted. Working with machinery can be very intimidating until one becomes accustomed to the way the machinery works. Training with an experienced operator is always a good idea and should be considered essential to safety.

11. **If you have a fear of difficult and dangerous situations**: Construction environments are often dangerous without appearing so at first. The simple idea of working in a trench for some utility work may seem safe, but it is actually one of the most dangerous situations possible. There is a danger of collapse and entrapment, but it is also possible to encounter a toxic and possibly fatal atmosphere in the trench. You may be able to work in a part of your chosen field that is less dangerous or difficult.

See also: Skill/Ability 10: Dealing with unfavorable conditions
Skill/Ability 12: Determination

Take Action: Identify your fears and identify how different professions may have many or few fearful situations for you. Decide whether you'd like to be doing work that challenges those fears, or work that keeps you safe from them. No judgment.

SKILL/ABILITY 10

Dealing with Unfavorable Conditions

Construction sites can be dirty, noisy, busy places. Work can start before dawn, and continue in rain, heat, or snow. Dealing with unfavorable conditions is part of the job.

1. **One of the key reasons Millennials do not pursue the construction trades as a career is because the conditions are often harsh.** If you are tough, strong, and determined, however, you can be very successful in construction.

2. **Early mornings!** All construction throughout the world usually starts early in the day. By the time most people arrive at work, construction crews have several hours work completed and are on their first break. This is real! Construction crews usually start early.

3. **Tradespeople are the ones who actually build the structures and install the plumbing electricity, and air-conditioning** that so many of us are accustomed to. Keep in mind that this means there is no electricity, air-conditioning, or plumbing until it is installed. Until these are provided then everything is on a temporary basis. Very often there are portable toilets, or generators running for electricity.

4. **Often the conditions are outside, in all kinds of weather.** Working outside can take a toll on your physical being. Sunburn, dehydration, and heat-related illnesses can be very dangerous in warmer places. In the winter, a cold day outside will tire you out very quickly. When it is extremely cold, frostbite and hypothermia become a big concern. It is said that you can dress for the weather, and you can. But you will need lots of rest and good diet to keep up with the pace of going outside to work every day. It's important to take care of yourself on the jobsite and off.

5. **NOISE!!! WHAT? NOISE!!!** All construction people will suffer some hearing loss at some point. It is very important to understand that jobsites are noisy. On large projects, the noise can be deafening. Everyone from new Millennials to experienced tradespeople should be using hearing protection. This is one of those unfavorable conditions which adds to the difficulty in construction work. In the long term, hearing loss is a reality, and Millennials coming into the trade should do everything they can to protect their hearing.

6. **There are many health hazards on jobsites**. These range from exposure to asbestos and silica to chemical hazards and many more. Millennials coming into the construction trades should take a Construction OSHA 30 Outreach Training class. Classes are designed to make the workers aware of jobsite hazards and safety issues. Workers need to be aware of what conditions can cause injury or death.

7. **Safety protection can go a long way.** From hearing protection and safety glasses to hard hats, gloves, and proper clothes, these protections can make unfavorable conditions somewhat more comfortable and much safer.

See also: **Skill/Ability 12:** Determination
Skill/Ability 13: Being Able to Learn from Mistakes
Success Tip 1: Safety First

Take Action: What kind of conditions are you willing to work in? Consider what is well within your comfort zone, at the edge of your comfort zone, and way into your discomfort zone!

Uncertainty in Employment

For many years people have looked at the construction trades as "temporary" employment. It was even very difficult in the past for construction people to obtain a credit card, because banks viewed tradespeople as seasonal employees with unreliable income. Over the years, this has changed considerably. Millennials no longer work for the same company for 30 years or more; preferences, technology, and companies shutting down or getting bought out all affect the stability of all jobs now.

1. **There is no such thing as a permanent construction job**; as a tradesman or tradeswoman, there is no guarantee to work for any company long term. Although you may like the company, and do very well with them, it is not unusual for a construction company to go out of business. Construction companies are often smaller type corporations, especially if they are trade specific, such as an electrical company or a carpentry company.

2. **As a tradesperson, you may work for multiple employers per year.** Even if you work with one company for a while, you may work at many different locations as time goes along. This is the nature of trades. As one project is completed, the workers will move on to the next jobsite and apply their skills to the success of that project.

3. **The only thing for sure in the construction trades is that sooner or later, all jobs will end.** There is a saying in construction: "We are always working ourselves out of a job!"

4. **On a personal finance level, it can be difficult to manage finances, knowing that the job can end in a very short time.** Careful financial planning becomes essential for trades workers. All too often a newer person finds out the hard way that the big check which seemed so steady can come to an abrupt end. Then payments on an expensive pickup truck or a high mortgage or rent can put you into a financially stressful situation. Budgeting becomes important.

5. **It always seems that when you have lots of money you have very little time, and when you have the time you have very little money.** It is hard to find a balance between the two. When someone is out of work, they have to be careful with spending. So how do you plan for a vacation or a home remodeling project? One of the best ideas is to set a personal budget based on ten months of employment year instead of a consistent twelve months of a year employment. By budgeting for ten months income, you build in some leeway in the event of an unexpected layoff or abrupt job conclusion. Abrupt job conclusions are not uncommon.

6. **The good news is that the trades have been evolving with technology throughout human existence, so they continue to be in demand.** Construction trades are very marketable skills. Just think about a plumber, for example. Everyone uses plumbing every day. Each day the plumber may be in a different location, performing a different task, but he is still doing plumbing. As the plumber learns and adapts to newer systems, he is able to maintain and keep pace with the demands for his services. The end result is that while no particular job site will be permanent, the trade is permanent. The demand for skilled labor in construction is a constant. As you learn and adapt to the newer systems, you ensure permanent employment.

7. **Tradespeople may not work for the same company for 30 years, but they work at the same *trade* for 30 years.** Very often,

smart, industrious, hard-working tradesmen and tradeswomen can do very well for themselves financially. So while job to job may not be steady employment, there is a demand for the skills and if one works hard at being a professional, there is a huge opportunity for a solid career in the construction fields.

See also: **Skill/Ability 10:** Dealing with unfavorable conditions
Skill/Ability 12: Determination
Success Tip 2: Stay tough

Take Action: What is your comfort level with periodic unemployment or working job to job. Some Millennials enjoy the change of pace and comfort of completing a job; others are terrified at being unemployed. What can you do to increase your comfort level, whether picking a work environment that meets your needs, increasing your savings, or something else?

SKILL/ABILITY 12

Determination

One of the key qualities of any successful person is that they work to accomplish the goals that they have set in their mind. For construction trades this is no different. You have to want to do the trade. You have to believe that it is in your best interests to work at the trade. It must be rewarding to perform the work, both in satisfaction of performing a skill, and also in terms of earning a living and being able to earn enough income to provide for a good living.

1. **Abraham Lincoln said, "Always bear in mind that your own resolution to succeed is more important than any other."** This is true of any occupation, but it is especially true in construction, because of the physical nature and trying conditions of the industry. You really have to want to do the work! You really have to want to make a career of it to be successful.

2. **Determination includes showing up to work *every single day*.** Missing work not only reflects poorly on you, it can overburden your colleagues and create unsafe conditions.

3. **As tradesmen and trades women improve their skills, they realize that they are of value to their employers.** They come to enjoy providing a valuable service for the wages they are being paid. Some construction foremen/forewomen have been paid large amounts of money to bring in projects on time. They are highly skilled and have great leadership ability, but in addition, they all have one thing in common: they are determined to get the job done. This is an essential trait among productive trades workers

4. **It is important to be resilient.** This means that you can bounce back after experiencing failure or a setback. We all have had negative things happen in our lives. Addressing problems, choosing to be positive, and moving forward will serve you well.

5. **Resourcefulness is also a part of determination.** In construction it's important to have the ability to stay with a task and see it through to the end, to be able to find creative ways to work around barriers, and to take satisfaction in accomplishing a task.

6. **Skilled and experienced tradespeople are confident in their abilities and their skills.** If you go to work with the idea of being happy to produce a quality product in a timely fashion, it is almost a matter of pride to be able to overcome any and all obstacles and challenges faced in the completion of a task or job as assigned.

7. **Be determined to be good at your trade,** and also to accomplish the tasks at hand.

8. **Determination also applies to studying in trade school, improving your skills in math or reading, overcoming fears, and working in difficult situations.** You **can** do this!

> **See also: Skill/Ability 9:** Ability to Overcome Fears
> **Skill/Ability 10:** Dealing with Unfavorable Conditions

Take Action: Set a challenging goal for yourself. Set a time limit. Determine all the steps necessary to accomplish the goal. Work at learning all that you need to know and become practiced at everything necessary to accomplish the goal. Work continually. Every day do something to bring you closer to that goal. Do not stop until you accomplish what it is you set out to do. Then realize that you have in you all that is required to be successful at a trade!

Being Able to Learn from Mistakes

Everyone makes mistakes! It's important to be able to learn from your mistakes, so you can keep making new mistakes instead of the same mistake over and over.

1. **We always want things to be perfect, and mistakes can feel awful.** Yes, there are consequences, but most mistakes can be fixed. Learning from your mistakes helps you improve.

2. **It's also important to learn from others' mistakes.** You don't have to personally make the mistake to have an opportunity to learn and grow. Observe others and pay attention to what goes right and to what goes wrong. If you see something done wrong, you can at least learn an example of something that you should not do.

3. **When you're in training and apprenticeship, don't let mistakes get you down.** It's part of learning! Just keep improving.

4. **A great way to learn from mistakes is to observe others.** If they are doing a task more efficiently or with greater ease then it is always good to learn that method, and if needed replace the way you previously did things with the new and better way. If this idea is applied over and again, it will help to improve skills quickly.

5. **On a large scale, trade instruction learned in school or a**

classroom setting is only the bare minimum and a beginning. The actual trade is learned in the field, hands-on and from experiences. Repetition is an excellent teacher, but the greatest teachers of all are adversity and mistakes!

6. **One way to improve and learn from mistakes is to ask yourself if the task being done is frustrating and difficult**. Smart Millennials are always looking for a simpler easier way to do something. If you are struggling to do something, you may be doing it wrong without even realizing it. Sometimes the mistake is not questioning the difficulty. By seeking an easier way, you improve and develop your skills. Fewer mistakes make for an easier workday and more production. Total satisfaction for both employer and worker is always a "win-win" situation.

See also: **Skill/Ability 7:** Logical thinking and planning
Skill/Ability 10: Dealing with unfavorable conditions
Skill/Ability 12: Determination

Take Action: Think back to when you made a big mistake. Now ponder what you learned from it. Even if you didn't learn something at the time, most likely there is something you can realize now that you learned. Keep in mind that you can and do learn from mistakes. Stay focused and keep improving!

Success Tips for Construction Workers

SUCCESS TIP 1

Safety First

Make no mistake: Construction work is dangerous! Safety is without a doubt the single most important concept on a construction jobsite. Safety consciousness has been a long slow process for many decades. Sadly, death and serious injury are still common in construction today. There are many initiatives to promote safety, beginning with regulations by the Occupational Safety and Health Administration (OSHA), but they still lag behind the issues that affect construction workers on a daily basis. Many workers and unions have fought for safety equipment and Personal Protective Equipment (PPE) to reduce injuries and fatalities on job sites. Following safety guidance helps keep you and others safe.

1. **Be advised: To Millennials entering into a construction career, this is a dangerous occupation!**

2. **Know the rules.** The Occupational Safety and Health Administration 29 CFR Part 5 (Federal Law 29 Code of Federal Regulations Part 5) states that employers shall furnish to each of their employees a place of employment which is free from recognized hazards that are causing or are likely to cause death or serious physical harm to their employees.

3. **Everyone going into construction should take an OSHA Outreach Training Class.** At the very minimum it should be an OSHA 10 Hour Outreach Training specifically targeted for construction trades.

4. **Pay attention in training about safety.** The information that is taught will be helpful later.

5. **Ensure you have appropriate safety equipment for the job,** and that your equipment is in good working order. Invest in safety equipment of sufficient quality and durability for the work if you can afford it.

6. **Stay sober.** Substance use and construction trades do not mix and can be extremely dangerous. Many construction employers fire employees who show up under the influence. Employers can legally require drug testing on a jobsite. They can require workers to be sober to maintain the safe operation of machinery and equipment. If you have a problem with drugs or alcohol, or see someone on your jobsite who does, seek help through your company's Employee Assistance Program, local hotlines, or local medical providers.

7. **Periodically inspect your safety equipment to ensure it is not damaged** with rips, tears, fraying, or other problems that could impact its effectiveness.

8. **Be aware of your environment to be alert for potential safety hazards for you and your coworkers.** The most common workplace safety violations are those related to falls, scaffolding, respiratory protection, ladders, and eye and face protection.

9. **Understand that the federal agency to oversee the safety of workers—Occupational Safety and Health Administration—was only created in 1970.** Prior to that, there were only guidelines protecting workers' safety. Many workers who came before you advocated for increased safety standards.

See also: Skill/Ability 7: Logical Thinking and Planning
Skill/Ability 10: Dealing with Unfavorable Conditions
Success Tip 10: Keep Equipment in Good Order

Take Action: There is no wage worth a permanent injury or death! Note that these next action steps are in **BOLD CAPITALIZED LETTERS.** This is the only place in this book that you will see that. It is emphasized because it is that important!

1. **TAKE AN OSHA SAFETY OUTREACH TRAINING CLASS.**

2. **TAKE AN OSHA 10 OUTREACH SAFETY TRAINING OR OSHA 30 OUTREACH SAFETY TRAINING FOR THE CONSTRUCTION INDUSTRY!!!**

3. **ONCE YOU HAVE COMPLETED AN OSHA SAFETY TRAINING CLASS, BE MINDFUL OF SAFETY AND SEEK SAFETY TRAINING FOR ALL ACTIVITIES YOU MAY BE INVOLVED WITH THROUGHOUT YOUR CAREER!!**

4. **PAY ATTENTION TO SAFETY AS IF YOUR LIFE DEPENDS ON IT ... BECAUSE YOUR LIFE DEPENDS ON IT!!!!**

Stay Tough

Construction tradespeople are tough. Not just physically tough, but also mentally tough to see a job through in demanding conditions. One of the reasons many former military do well in construction is because of their training and experience from their military service, which translates well into the construction field. Toughness comes from the discipline to be able to continue even when you're tired, conditions are difficult, and you just want to go home.

1. **Recognize the toughness you already have**: You chose this profession, you enrolled in school, you worked your way through school, and you've already worked on job sites. Remember anything else that you've chosen, persisted at, and been successful in. You can do this.

2. **Construction starts early**. Be proud of your ability to rise up early and make your way in the world.

3. **Notice when you are doing something that requires toughness**, whether it's working in challenging weather conditions, working a long shift to meet a deadline, or doing precision work that requires your complete attention. Notice also that dealing with setbacks, pushing yourself to your limits, dealing with difficult people, and making difficult decisions are all ways of being tough as well.

4. **Remember that toughness does not mean disrespectful or crude**. Truly tough people don't put others down, call people names, or bully others.

5. **We all have moments when we don't feel very tough.** Remind yourself that we can keep going even when we don't feel great.

6. **Encourage your coworkers when you see them being tough** and find people on and off the job who can encourage you.

7. **Know that because you can handle all of the conditions and challenges of the construction world, you are very strong person indeed.** It is good to reassure yourself of your own strength.

> **See also: Skill/Ability 9:** Ability to Overcome Fears
> **Skill/Ability 10:** Dealing with Unfavorable Conditions
> **Skill/Ability 12:** Determination

Take Action: Think about a time when you saw a project or a difficult situation through to the end. Regardless of the difficulties or the hardship faced, there was a great sense of accomplishment when the goal was achieved. Bring that same determination to a jobsite. Be proud of having the toughness to get the job done.

Build on Your Strengths

The construction trades have a variety of occupations, from carpenters to surveyors and from ironworkers to telecommunications technicians. If you would like to be involved in this industry, it's important to identify your strengths, so that you can choose an occupation that helps you build on your strengths. If you are an organized person you may lean towards the management side of the industry, and if you are more mechanical you may lean more towards the machinist or engineering related trades. Look at what you are good at, and what you could like to do in order to develop a long-term career in the industry.

1. **Many Millennials choose construction because of the variety of the projects and the challenges of being part of the process to build something physical.** Often people choose construction because they do not want to be "behind a desk someplace."

2. **Going into the construction field is a choice.** If you look at it as you have to because you have no other choice, then this may not be for you.

3. **If you're not sure what your strengths are, ask** a mentor or instructor or supervisor you trust and who knows you well to help you identify your strengths.

4. **It's not all about size.** Within the construction trades, there's always an opportunity to maximize your strengths. For example,

if you are not physically very strong, you may be interested in finish work, operating equipment, or engineering-related work compared to activities that involve lifting heavy objects.

5. **Pay attention to what others are doing on worksites** so you can get an understanding of the many different opportunities within your trade – or even within other trades!

6. **When you find people with strengths similar to yours**, ask them what they do and how they build on their strengths. We all bring different strengths to the table. Identify what your strengths are, and work to shine in those areas.

See also: Skill/Ability 1: Physical ability
Skill/Ability 2: Mechanical ability
Skill/Ability 9: Ability to overcome fears

Take Action: What are your strengths? Really, take some time and think about what you do well, what you like to do, and how these strengths can translate into the construction trades.

SUCCESS TIP 4

Be on Time Every Day

Trades workers are usually paid by the hour, so from the very beginning, the agreement between employer and employee is based in a measurement of time: the hour. That assumption of course, is based on trust that the employee will actually perform work for the entire hour, and that the employer will in fact pay the employee for the time and labor expended. This seems like a fair arrangement. For many jobs this is standard.

1. **Being on time can mean different things to different people**: being at the jobsite, being at work and unpacking, or being onsite and ready to go. We suggest being onsite and ready to go.

2. **Construction companies run tight crews with very little room for error.** Very often members of a crew are depending on each other to show up so that they can accomplish the work at hand. When one member fails to show, it impacts the production of the entire crew. It's important that you go to work every single day. Missing days without notice or reason makes you look irresponsible, inconveniences your boss and colleagues, and can create unsafe working conditions. On time. Every single day.

3. **If you have a timecard and have to clock in and out, make sure you understand the rules** around how early you can clock in, how many hours you can work, and what activities are officially "on the clock" and "off the clock."

4. **If you need to get coffee to start your morning**, make sure that happens before you're scheduled to start work.

5. **Many times, work can't start until everyone is present to get the initial instructions and assignments.** By being on time, you won't delay the work for anyone. You help the crew to move forward and be productive by taking personal responsibility and being on time.

6. **In construction, time goes very fast.** It is always important to be efficient and productive with the time allotted. Not just to make a profit for the boss, but also to improve your skills and finish the project.

7. **In certain areas, workers may have to travel to the actual work-site after they arrive at the work location.** In the case of those who work on the water, they may travel by boat to the actual work location, and they are governed by the tides while working on their projects. Because of the tides, the boats will leave to start work at very specific times. If you are late, you literally miss the boat!! You also lose a day's pay.

See also: **Success Tip 7:** Value the Success of the Overall Project
Success Tip 8: Be Honest and Earn Trust

Take Action: What do you need to ensure you're on time every single day? What do you anticipate will get in your way? Late bus, broken alarm clock, late nights that make it hard to wake up? Anticipate these problems and deal with them.

Show up Prepared to Work

Being prepared to do a job is one of the most important traits that a trades person brings to a job or to a company. Being prepared includes being dressed properly, having the right tools, the right attitude and the knowledge of resources that can be drawn on when needed.

1. **A common construction saying is: "If you don't have it, you ain't got it!"** This means that if you didn't bring something with you, you will not have it available to you at all. If you forget to bring warm clothes in the winter, you will be cold. If you are don't bring water, you will get thirsty. If you forget your tools, you will not be able to perform your job. The trades can be harsh; going to work in construction forces you to be self-reliant. You learn to take care of yourself because everyone else is busy taking care of themselves.

2. **As much as possible, identify what gear is needed before showing up.** Wear proper gear to do the work, including gloves and boots.

3. **Bring proper gear for the weather,** including if it rains or snows.

4. **Bring appropriate tools, including safety gear,** as needed for the job.

5. **Ensure you have a plan for food and water to drink.** Lunch for purchase is not always readily available in construction. Keep in

mind that you get limited time for lunch. Fast food and delicatessens might be far away. It is a good idea to bring your own lunch from home.

6. **Working hard will make you thirsty.** Always bring water, just in case it is not provided on the jobsite or is hard to get to.

7. **Show up sober.** Substance use and construction trades do not mix and can be extremely dangerous. Many construction employers fire employees who show up under the influence. Employers can legally require drug testing on a jobsite. They can require workers to be sober to maintain the safe operation of machinery and equipment. If you have a problem with drugs or alcohol, or see someone on your jobsite who does, seek help through your company's Employee Assistance Program, local hotlines, or local medical providers.

8. **Many times, you have to carry everything you bring with you,** so be prepared, but don't bring too much gear.

9. **Be diligent in your preparations.** Often at a jobsite, there are no services available. There is no delicatessen to buy food or water, and there is no shelter in places if you are working outside. All of this needs to be taken into consideration when going out to work.

See also: Skill/Ability 12: Determination
Success Tip 2: Stay Tough
Success Tip 5: Show up Prepared to Work

Take Action: What do you need to bring to your jobsite or training program? Make a list and keep it by the door. Update the list as your site or program changes.

SUCCESS TIP 6

Show up with a Positive Attitude

"Attitude is everything, and everything is attitude." Companies focus on accomplishing their jobs. The company is bound to fulfill their contract, which is why they are called contractors. Companies place a high value on those who are motivated to help them accomplish their goal of fulfilling the contract.

1. **No one likes a sourpuss** (except maybe Grumpy Cat). A positive attitude will get you far!

2. **Being willing to work** – with a positive attitude — makes it more likely people will want to work with you.

3. **Show up sober.** Substance use and construction trades do not mix and can be extremely dangerous. Many construction employers fire employees who show up under the influence. Employers can legally require drug testing on a jobsite. They can require workers to be sober to maintain the safe operation of machinery and equipment. If you have a problem with drugs or alcohol, or see someone on your jobsite who does, seek help through your company's Employee Assistance Program, local hotlines, or local medical providers.

4. **Good, funny, clean jokes can get help lighten the mood on a job** and demonstrate your good attitude to others. Keep in mind "risky" jokes can get you kicked off the job site.

5. **Remember, there is so much to learn**. Approaching each day with an attitude of curiosity can serve you well.

6. **Being willing to do whatever you are assigned** can help build your reputation as someone who is reliable.

> **See also:** **Skill/Ability 10:** Dealing with unfavorable conditions
> **Skill/Ability 12:** Determination
> **Success Tip 2:** Stay Tough

Take Action: Even if you're not a morning person, you can train yourself to be positive (or at least more positive) in the morning. What will it take for you: Coffee? A good breakfast? A good joke? Listing to your favorite music on your way to work? Figure it out, and make it happen!

SUCCESS TIP 7

Value the Success of the Project

Often construction workers are hired on a temporary basis. They are hired just for skills and abilities. They are not necessarily emotionally invested in a project as much as an engineer or an architect, because those roles see the project from concept to completion. While trades-workers may not have an emotional bond to the project, they do have a pride in the work that they do to see to it that the work is done properly and in a timely fashion. This is their contribution to the success of the overall project.

1. **Although your part of the overall project may be small,** you're contributing to something exciting, whether it's a piece of furniture, a bridge, or a building. Doing your part well can help keep you motivated.

2. **Bear in mind that while your work may seem insignificant at first, each and every task must be done in order to accomplish the project as a whole.** The decisions are made at the higher levels as to what needs to be done. Because there is usually a lot of experience behind those who are making the decisions, the decisions are well thought through.

3. **Crew leaders on the jobsite will sometimes look for suggestions among their crew members.** They will value those who have tried to do their jobs well. If looking for suggestions, the leadership will always default to the more proficient and

knowledgeable crew members who are usually the most experienced, sometimes with more training and experience than those running the crew. They will trust them because they know that those tradesmen will always do their jobs well. That is how it is. That is how they roll.

4. **Being willing to help colleagues can contribute to the success of the overall project.** If you see someone struggling or unsure, help them out.

5. **Your contributions to the larger project will make you proud!** Being part of a building or bridge is something you can share with friends and family forever.

6. **It is valuable to take pride in the work you do.** Knowing that you did the work to the set standard means that you have made a significant contribution to the overall success of a project. All tasks on a job must be complete, even if they seem insignificant at first.

See also: **Skill/Ability 12:** Determination
Skill/Ability 13: Being able to learn from mistakes
Success Tip 4: Be on Time, Every Day
Success Tip 9: Focus on getting stuff done

Take Action: Think back to when you've worked in group project: Did you tend to get lost in your part, or focus on your contribution to the whole project? Changing your perspective can change your life.

SUCCESS TIP 8

Be Honest and Earn Trust

No one likes to be cheated or lied to. Everyone wants to be treated fairly both as person and in general business. The same is true in construction. But in construction, the gloves come off. If you wrong someone, they never forget. If you lie to someone, you're forever a liar. If you are not trustworthy, most of the time you will be called out on it – and you might get fired.

1. **Say what you mean and mean what you say.** Building a reputation as a straightforward worker can serve you well.

2. **When someone gets the reputation of being untrustworthy,** it is impossible to shake off that reputation. While it may never be said outright, the reputation holds forever.

3. **Do not steal or make jokes about stealing.** Bosses take this seriously and it could cost you your job, even if you didn't steal anything.

4. **It is ok to say: "I have never done that before"** or "I have no experience at this" or "I just don't know." That's better than lying and being found out.

5. **Everything about how you present yourself,** including showing up on time and prepared, talking respectfully to others, and doing a good job, demonstrates your trustworthiness.

6. **Understand the rules of timecards and hours worked** so you don't steal time. Do honest work for honest pay.

7. **If you don't know an answer to a question, say so.** If you try to fake it, you'll get caught and others won't trust you.

8. **No one likes a liar;** If you cannot be trusted they will not keep you. It is that simple. Remember you are trying to make a living, and you need to become a trusted member of a crew to stay employed!

9. **No one ever forgets who has done them wrong in the past.** Once you are labelled as untrustworthy, it is almost impossible to shake that stigma.

10. **Old idea that's still true: Those who are untruthful in small things will be untruthful in larger matters too.**

See also: **Skill/Ability 7:** Logical Thinking and Planning
Skill/Ability 10: Dealing with Unfavorable Conditions
Success Tip 4: Be on Time, Every Day
Success Tip 14: Show Solidarity

Take Action: This one's easy. Tell the truth. Be honest. Guard your reputation and your character by doing the right thing.

Focus on Getting Stuff Done

There's one thing for sure in construction: If you produce lots of high-quality work on a regular basis, they will keep you. They may not like you, but they will keep you. When the project comes to an end, they may lay you off, but at least you know you can go back and ask for a job with that company again in the future.

1. **A productive attitude does not mean that you are the most productive person at work; it means that you are working with production in mind.** You are focused on contributing your effort to help produce an end result. This differs greatly from someone who simply goes to work not caring whether a task is completed or not. A person who is simply filling time is not as valuable as someone who contributes to the overall success of a project.

2. **Pay attention to others at the worksite to identify the pace of work.** Ensure you're at least above average in your pace.

3. **Those who know their trade well will perform at a good rate, and usually they make it look easy.** Make sure you know your skills and study up or ask others if you are not sure.

4. **Most of the people who stay in the trades want to see their work done right,** and they want to see the job through to a successful conclusion. They take pride in the work they do.

5. **In the course of any job, there are always menial tasks that**

need to be tended to. Often they are of lesser priority and left undone. As a new person you can always keep busy and render a service by doing some of these tasks whenever there is "downtime." You can always grab a broom and sweep the floor, you can clean the workspace, or put unused tools away. This is one way of how to show a positive, productive attitude.

See also: Success Tip 1: Safety First
Success Tip 7: Value the Success of the Project
Success Tip 11: Take Initiative

Take Action: Look online for videos of tradespeople doing different jobs. Look at anything from steel erection to finish painting. There are many examples of beautiful work in all fields, and you can see high levels of skill being performed. Take note of the way the work is being done. Look at the end results. If the tradespeople show their faces, you will be able to observe a satisfaction in their expressions. They have taken the time to make a video of the work. Really, they are showing off! They are proud of their accomplishment. It shows off not just the work, but also who they are, and how they have mastered the skills that they are showing in the video.

SUCCESS TIP 10

Keep Equipment in Good Order

Tradespeople know their success in construction skills is because they can count on two things: confidence in their skills, and confidence in their tools and equipment. Buying good-quality tools and equipment pays in the long run. Because there is a high price for good quality, then that investment should be protected. The way to get the most value for that investment is to take care of the tools and equipment to ensure a long life of service from those items. When tools and equipment are well cared for, their value increases. The value to the tradesperson comes in the satisfaction of being able to complete a task with high performance. The value of the equipment to the company owner is reflected in the bottom line of cost effectiveness.

1. **Pay attention in trade school about how to maintain and repair equipment.** This information will come in handy later.

2. **Buy good-quality tools and take care of them.** The good-quality equipment is reliable and will serve for a lifetime.

3. **Treat equipment you use like it's your own and take care of it accordingly.** Care for the company tools the same way that you would care for your own personal tools.

4. **Mark your own personal equipment and tools with your own "mark"** so that you know what's yours and what's not.

5. **Clean equipment and tools properly** when you are finished with them.

6. **If you see a potential problem with equipment, alert your supervisor.** If there is a safety issue, alert your supervisor immediately.

See also: **Success Tip 5:** Show up Prepared to Work
Success Tip 7: Value the Success of the Project
Success Tip 11: Take Initiative

Take Action: Choose a trade to investigate. Look up the costs of various tools and equipment use for that trade. Immediately you will notice that there are less expensive tools and equipment and then there are more expensive tools and equipment. The difference between the two varieties is almost always quality. Ask yourself why there would be such a difference in price. Tools that are of higher quality will always perform and outlast the tools of lesser quality. Many times, the higher better-quality equipment will turn out to be less expensive in the long run, because they will not need to be replaced as often as the seemingly bargain priced tools or equipment.

Take Initiative

A willingness to work is clearly demonstrated by taking initiative to work without someone having to specifically tell you each and every single item that has to be done. Workers who continue to perform without immediate supervision on every detail will be more valuable to a company because they show leadership. They might be able to manage projects for a company, eventually becoming a foreman or supervisor.

1. **We all start somewhere.** When you're new to a job or site, pay attention to what is going on around you. Try to relate what you're seeing others do on the job to what you learned in school or what you already know how to do. Identify gaps and ask questions.

2. **Identify the steps of projects,** so you understand both your steps in the process, and also how you can help those with steps before and after yours.

3. **Once you learn how to do something, help those with less experience.** Not only can it be very satisfying to teach others, you might learn something new yourself!

4. **Observe the workplace to identify what the best way is to take initiative.** Sometimes bosses prefer workers to raise their hands to volunteer; others prefer you approach them individually. Figure out what happens at your workplace, and start taking initiative.

5. **Taking initiative with a positive attitude will help get you**

noticed as reliable, thoughtful, and hardworking. That will serve you well.

6. **Work for someone else the way you would like to have things done for you.** Do this all the time, not just sometimes.

7. **Make it a habit of delivering quality as if you were doing things for your own benefit.** In the overall picture, you will improve as a worker and it will be appreciated by those that hire you. In addition, you will find more satisfaction in the work you do.

See also: **Success Tip 5:** Show up Prepared to Work
Success Tip 6: Show up with a Positive Attitude
Success Tip 7: Value the Success of the Overall Project

Take Action: Observe who on your jobsite is highly respected, whether another worker or a supervisor. Identify what they do (or don't do) that leads people to respect them, including taking initiative to do what's right.

SUCCESS TIP 12

Become an Expert at Your Trade

Some Millennials work in the trades because they feel it is just a job for money. This is not always to their benefit. The ones who take an interest to become a good or even excellent trades person always find a more rewarding higher paying career than those who are not as dedicated.

1. **You joined the construction trades for a reason**. Even if that reason was only to make money, you might as well learn something while you're here! See what opportunities there are to learn in every project and job.

2. **Curiosity is a positive value in the construction trades**. Being open to learn new things is key to advancing and performing well.

3. **Identify what advancement opportunities there are in your trade,** such as from apprentice to journeyman. Figure out what the steps are to advance and start learning!

4. **Talk to more senior people** to identify what they wish they'd learned earlier.

5. **Ask for feedback when appropriate** on how you completed a project or where you could improve.

6. **A great way to become interested in the trade is to find out**

about new methods and technologies, learning about what is new and coming into the trade. Keep up with the latest tools and techniques being developed.

7. **Invest in yourself as a tradesperson.** Look to learn all you can about your field. Become familiar with the language of the trade that interests them, including technologies, work processes, tools, and materials. Use the internet: the more you look, the more you will find. All of this will benefit you as a worker and as a person.

8. **For those already involved in the trades, education always pays.** The more well-rounded, skilled and knowledgeable you are, the more you will find work. The tradespeople who make the highest wages and compensation are always well trained and educated in their field

See also: Skill/Ability 12: Determination
Skill/Ability 13: Being able to learn from mistakes
Success Tip 5: Show up with a Positive Attitude

Take Action: Start now!! Someone who is interested in a career in construction will be well served to learn from all sources available about the trade that interests them. Even before you come into a trade, start to learn all you can about that trade.

SUCCESS TIP 13

Always Be
Learning More

The construction trades are always changing. Those who can grasp new ideas and learn how to apply them hold a key advantage over those who do not learn. As a Millennial, you have an advantage over others, because you are familiar with learning and adapting to new technologies. Tradespeople with the ability to understand and work with new technologies will find themselves with a long and rewarding career.

1. **Curiosity is a positive value in the construction trades.** Being open to learn new things is key to advancing and performing well.

2. **Note that there are many different ways to learn**: watching others, reading a manual, watching a video online, taking a course, and more. Find out what ways of learning work well for you.

3. **There are also many sources of new technologies**: unions and guilds sometimes offer training, equipment manufacturers will share new technologies (and try to sell them to you!), and your employer may offer training on new technologies as well.

4. **Identify people who have more experience than you and ask them** about what's important to know and how they learned on the job. These tips could be helpful for your own learning process.

5. **If you get frustrated learning something new, that's okay** – it's part of the process. Take a deep breath, slow down, and try again.

6. **Set goals for yourself or identify problems to solve.** Whether you're a goal-reacher or problem-solver, you've just set yourself up to learn something new!

See also: Success Tip 3: Build on Your Strengths
Success Tip 11: Take Initiative
Success Tip 12: Become an Expert at Your Trade

Take Action: Do some research into how new technologies are being introduced in your field of interest and learn those systems. This may not pay at first, but it will be a great investment for a long-term career.

Show Solidarity

Solidarity is an important value in construction trades. It is the sense that each person should look out for the interests of all, including co-workers and the company.

1. **Work for the company, not just the wages.** Know that the tradespeople and others who make up the core group in the field play a key role in defining the company. They become the face of the company. These are also the ones that make the high wages and benefits.

2. **Your company includes more than just tradesworkers**: the office administrators, dispatchers, supply clerks, field crews, and managers are all on the same team as you. If you want this to be a career, not just another job, have solidarity with **all** of your coworkers.

3. **Similarly, solidarity means not being aggressive, sexist, racist or homophobic** toward coworkers. Treat all your coworkers with respect.

4. **Many construction companies put a high value on those that have interest in the success of the company, and in turn, they take care of their workers.** This is sometimes called "riding for the brand." Companies and coworkers can always tell which employees are dedicated and who is simply "here for the income and not the outcome." Be someone who can be counted on. Always be professional.

5. Learn from the highest quality tradespeople and try to be as good as they are.

6. Realize that being with a company for a long time helps to improve your "hands on" skills and increase your professionalism.

7. Consistency is to the benefit of the individual as well as to the company.

See also: Skill/Ability 10: Dealing with Unfavorable Conditions
Success Tip 2: Stay Tough

Take Action: Identify people you know who have made a good career for themselves in construction. Spend some time with them. Ask what they do for work. Ask what they like and what they don't like about their jobs and careers. See how long they have stayed with a company. You may be surprised by the level of skills and education they have received along the way. Pay attention to the attitude and professionalism that they project about their position and their career. See if this is something you may like to do for yourself.

Common Challenges and Opportunities in the Trades

CHALLENGE/OPPORTUNITY 1

Should I Join a Union?

You might have heard a lot about unions in the news, and the odds are that when you join the construction trades, you will hear even more. Here are some things to consider when you are deciding whether to join a union:

1. **About 13% of construction workers are represented by unions,** compared to about 7% of all workers. Sometimes specific construction jobs are only available to union members; other job sites do not accept union members. You'll need to check out the situation in your profession and area.

2. **If you enroll in a union-sponsored apprenticeship,** you get training for a trade, a salary while training, and eligibility for union membership. Aligning with a union can be a great way to get started at a higher pay rate than you would otherwise.

3. **Union membership has been declining across the U.S. over the past 50 years** for several reasons: Unions are increasingly viewed as not relevant, because they often can't protect workers from layoffs. Union labor is more expensive than non-union labor in the U.S. or any labor elsewhere in the world (where many other countries don't have worker protections and benefits like in the U.S.). Unions can also be perceived as politically non-neutral, or as inefficient or corrupt.

4. **Despite these negative perceptions, union members earn better wages and benefits than workers who aren't union**

members. On average, union workers' wages are 28 percent higher than their nonunion counterparts; in construction, union workers' wages are a full 44% higher than nonunion counterparts.

5. **Labor unions also give workers the power to negotiate** for more favorable working conditions and other benefits—such as salary, health insurance, pensions, or training—through collective bargaining.

6. **Union members typically pay union dues**. Dues are the cost of membership to be in the union; they are used to fund the various activities the union conducts, including organizing, administration, attorney costs, training, and staff to assist in negotiations, grievances, and arbitration.

7. **Trade unions are champions of equal rights and equal pay**. They also fight discrimination, including standing up for union members if they are being discriminated against or treated unfairly.

See also: Success Tip 8: Be Honest and Earn Trust
Success Tip 12: Become an Expert at Your Trade
Success Tip 14: Show Solidarity

Take Action: Look online or in the library for information about the unions that typically represent your trade. Ask around to find out more information about what it is like to work as a union member and as a non-union member. That way, you'll have more information to make an informed choice.

CHALLENGE/OPPORTUNITY 2

Being a Woman in the Trades

Construction is without question a male-dominated field. The proportion of women of all ages involved in the construction trades is around 9%, but only about 5% of Millennial women work in construction. Working in the trades as a woman is an excellent way to do skilled work for good pay.

1. **Many construction companies have a high demand for skilled labor.** You are likely to be hired on the basis of your qualifications, skills and abilities rather than race or skin color. This presents an opportunity for Millennials of any race or background to earn a living in the construction trades.

2. **It is more important to be a good tradesperson than it is to prove you just as strong as someone who is bigger or stronger than you!** It is not about physical strength, it is about consistently strong performance. Employers want the jobs done. They are not as concerned about the appearance of the person who accomplished the work.

3. **There are jerks everywhere, including in construction.** No one should tolerate discrimination or sexual harassment. If you believe you are being treated unfairly, consider discussing the problem with your boss, your organization's human resources department, or your union representative.

4. **Very few women, if any, enter the construction trades looking for romance.** The vast majority of tradeswomen are working at earning a living just like everyone else. Male and female tradespeople should always keep this in mind. When at work, it is about getting the job done and performing the skills of your trade. This usually has nothing to do with relationships.

5. **There are many national organizations to support women in the trades,** including National Association of Women in Construction (NAWIC), National Association of Black Women in Construction (NABWIC), Nontraditional Employment for Women (NEW), Professional Women in Construction (PWC), Apprenticeship & Nontraditional Employment for Women (ANEW), National Association for Women in Masonry (NAWM), National Institute for Women in Trades, Technology & Science (iWiTTS), StartZone, Sisters in the Building Trades, The National Taskforce on Tradeswomen's Issues, Policy Group on Tradeswomen's Issues (PGTI), Tradeswomen, Inc, Federation of Women Contractors (FWC), Women in Non Traditional Employment Roles (WINTER), Coalition of Labor Union Women (CLUW), Minority and Female Skill Trades Association, Women in HVACR, and Women in Construction. There are also local organizations, such as Chicago Women in the Trades (CWIT), Missouri Women in Trades (MoWIT), Oregon Tradeswomen, Vermont Works for Women, Washington Women in Trades, and Central Ohio Women in the Trades, to name a few. Find other women who do what you do, and you will feel more confident in your work.

6. **Most cities and states have a contractor list for minority and women-owned businesses.** Working for a construction company that is woman-owned and operated may be a different experience.

7. **Physical ability and physical strength are not the only skills** required in the construction trades; not all construction jobs are only for those with high physical strength.

8. **If you are mechanically minded and are willing to work hard,** you can do very well in the construction trades, including advancing to be managers and leaders.

9. **Millennials who do well in the trades focus on learning the trade and are willing to do all that is necessary to perform the tasks assigned.** They focus on getting the job done and done well.

10. **No one should ever sell themselves short on being "tough enough."** Remember toughness comes from being able to focus and perform under difficult conditions. When you establish yourself as a solid, reliable, productive worker, your personal characteristics will have almost no bearing on your employment.

See also: Skill/Ability 12: Determination
Success Tip 2: Stay Tough
Success Tip 3: Build on your Strengths
Challenge/Opportunity 7: Juggling work and parenting
Challenge/Opportunity 15: Colleagues are aggressive, racist, sexist, or homophobic.

Take Action: Check out programs that support and encourage women in the trades. Reach out to trade unions for training, because they are likely to provide more support than going it on your own. If you're not sure if you want to pursue a career in construction, check out adult education programs in woodworking or home repair to see if you enjoy it.

Being an Underrepresented Racial Minority in the Trades

In the past, construction trades have been viewed as a place where individuals who are not white men were not welcome. Here in the new millennium, minorities are increasingly represented in the construction trades. Currently Latinx individuals are 30% of construction workers and African Americans are 17% of construction workers.

1. **Many construction companies have a high demand for skilled labor.** You are likely to be hired on the basis of your qualifications, skills and abilities rather than race or skin color. This presents an opportunity for Millennials of any race or background to earn a living in the construction trades.

2. **There are jerks everywhere, including in construction.** No one should tolerate discrimination or harassment. If you believe are being treated unfairly, check it out and seek support.

3. **Most trades do not require a college degree.** Training programs, especially through unions and guilds, have embraced fairness and equality and often reserve places in their apprenticeship programs for Millennials of color or from disadvantaged communities. This is a good place to look to get started.

4. **Look into apprenticeship programs so you can get started training – and get started bringing in an income.** Look at the U.S. Department of Labor website and other resources listed in this book. Once you find a pre-apprenticeship or an apprenticeship program, apply and get started.

5. **There are many organizations to support minorities in the trades,** including National Hispanic Construction Association, National Association for Black Women in Construction (NABWIC), National Black Contractors Association, National Association of Minority Contractors (NAMC), the Chinese Staff and Worker's Association (CSWA), StartZone, and Minority and Female Skill Trades Association. Find other people like you who do what you do, and you will feel more confident in your work.

6. **Most cities and states have a contractor list for minority and women-owned businesses.** Working for a construction company that is minority-owned and operated may be a different experience. Most government-funded projects have set-asides to facilitate minority-owned business participation.

7. **Accept that your workmates are all working for the same reason: to earn a living.**

8. **If you are mechanically minded and are willing to work hard,** you can do very well in the construction trades, including advancing to be a manager and leader.

9. **Millennials who do well in the trades focus on learning the trade and are willing to do all that is necessary to perform the tasks assigned.** They focus on getting the job done and done well.

10. **No one should ever sell themselves short on being "tough enough."** Remember toughness comes from being able to focus and perform under difficult conditions. When you establish

yourself as a solid, reliable, productive worker, your personal characteristics will have almost no bearing on your employment.

See also: Skill/Ability 12: Determination
Success Tip 2: Stay Tough
Success Tip 3: Build on your Strengths
Challenge/Opportunity 15: Colleagues are aggressive, racist, sexist, or homophobic.

Take Action: If you don't feel supported, seek specialty organizations where you can find solidarity. There is a list in the For Further Reading section.

CHALLENGE/OPPORTUNITY 4

Being an LGBTQ person in the trades

Although still lagging behind civil rights for other aspects of identity, the U.S. Supreme Court affirmed in 2020 that companies cannot fire someone for being lesbian, gay, bisexual, transgender or queer (LGBTQ). This is a huge achievement!

1. **Many construction companies have a high demand for skilled labor.** You are likely to be hired on the basis of your qualifications, skills and abilities rather than other characteristics. This presents an opportunity for all Millennials to earn a living in the construction trades.

2. **There are jerks everywhere, including in construction.** No one should tolerate discrimination or sexual harassment. If you believe are being treated unfairly, check it out and seek support.

3. **There are some organizations to support LGBTQ individuals in the trades.** Pride at Work and Build Out Alliance organize mutual support between organized labor and the LGBTQ community, and there may be local organizations in your city or state. Find other LGBTQ individuals who do what you do, and you will feel more confident in your work.

4. **The National LBBT Chamber of Commerce (NGLCC) certifies LGBTQ-owned businesses and many cities also have LGBTQ Chambers of Commerce.** You can talk to them about

LGBTQ-friendly construction businesses. Working for a construction company that is LGBTQ-owned and operated may be a different experience.

5. **If you are mechanically minded and are willing to work hard,** you can do very well in the construction trades, including advancing to be a manager and leader.

6. Millennials **who do well in the trades focus on learning the trade and are willing to do all that is necessary to perform the tasks assigned.** They focus on getting the job done and done well.

7. **No one should ever sell themselves short on being "tough enough."** Remember toughness comes from being able to focus and perform under difficult conditions.

> **See also:** **Skill/Ability 12:** Determination
> **Success Tip 2:** Stay Tough
> **Success Tip 3:** Build on your Strengths
> **Challenge/Opportunity 15:** Colleagues are aggressive, racist, sexist, or homophobic.

Take Action: Check out programs that support and encourage LGBTQ individuals in the trades. Also reach out to trade unions for training, because they are likely to provide more support than going it on your own.

Being a Veteran in the Trades

Many Millennials join the construction trades from the military. Some Millennials learned mechanical skills in the military, which often directly translate to construction. Other times, the qualities and values learned in the military—including working in challenging conditions, determination, toughness, and solidarity—are a good fit for the construction trades. As a result, they can earn a high wage doing something they have learned and provide a good way of life for themselves and their family.

1. **Veterans are usually highly respected in the trades.** Many tradesmen and women are also veterans and so embrace their fellow veterans as they enter into construction. Because of the large number of veterans in the trades there is an advantage for veterans in building good relationships and finding mentors

2. **If you are still in the military or recently discharged from the military, check out programs in the military to help service-members transition out of military service.** The programs Helmets to Hardhats, Troops to Trades, and Military Skilled Trades are specifically designed to help place military personal who are veterans, or coming off active duty, into the construction trades. They are able to join the trade unions with little or no difficulty and almost no waiting time. This is a huge advantage for the military personnel looking to transition into civilian life, with a good career opportunity.

3. **Discipline learned in the military translates well into the trades.**

4. **The education and training received in the military usually blends well in the construction industry,** with similar values and work environments.

5. **One challenge for veterans is the adjustment from working under military rules to civilian lifestyles.** Remember what you learned in the military can be extraordinarily helpful as a civilian, as good habits such as being on time and putting in long hours will benefit you as you find your way in the trades.

See also: Skill/Ability 12: Determination
Success Tip 2: Stay Tough
Success Tip 3: Build on your Strengths

Take Action: Regardless of where you are in your military or construction careers, check out Helmets to Hardhats, Troops to Trades, or Military Skilled Trades. They offer resources, support, and answers.

CHALLENGE/OPPORTUNITY 6

Being a Person with a Criminal Conviction in the Trades

Unfortunately, there are Millennials who have, for whatever reason, a criminal conviction in their personal history. This is often a stumbling block to Millennials trying to leave their past behind and start over after having "paid their debt to society." Very often, there are few job opportunities for someone with a criminal record. One of the few places where someone can leave their past behind is in the construction trades: you are hired based on your preparedness, ability, and willingness to work. It is possible to move forward. If this applies to you, keep reading!

1. **After having served time for an offense, it is often difficult to leave the past behind.** The situation may feel hopeless and there is often a temptation to return to the same behaviors and activities that led to the jail sentence in the first place. To move forward, decide not to return to the former lifestyle. Decide that that is in the past.

2. **Get the necessary education.** Begin immediately to complete a basic education for either a high school diploma or a GED (General Education Diploma)/high school equivalency diploma, if necessary. This can often be completed while incarcerated or immediately afterward.

3. **Look into apprenticeship programs as soon as possible so you can get started training—and get started bringing in an**

income. Look at the U.S. Department of Labor website and other resources listed in this book. Once you find a pre-apprenticeship or an apprenticeship program, apply and get started.

4. **If possible, take basic fundamental skills education classes for the construction trades in general,** as well as for the particular trade you are interested in. This will increase your skills and opportunities.

5. **Work hard at becoming excellent in your coursework and at your trade's skills.** Study and practice!

6. **As soon as possible, take the credentials you have and begin looking for work.**

7. **Once you are employed, dedicate yourself to becoming excellent at your trade.** Realize that an apprenticeship is a starting point. You will have to work at the trades for a few years before it really starts to pay back the time invested in apprenticeship. However, you will still be earning a living while learning the trade.

8. **Plan out what you want your life to be.** Decide what kind of life you want and work to achieve that life.

9. **Invest in yourself and learn all you can.** Look for opportunities that will benefit you personally and will help you achieve your goals.

10. **Remember: success is possible, but it will require hard work and focus on planning for a better future.**

> **See also:** **Skill/Ability 12:** Determination
> **Success Tip 2:** Stay Tough
> **Success Tip 3:** Build on your Strengths

Take Action: Consider what you want your life to look like. Seek support to help you get there. You got this!

CHALLENGE/OPPORTUNITY 7

Juggling work and parenting

Being a parent in the trades is a huge challenge! Construction jobs are not "9 to 5." Very often the schedules are governed by the demands of the job at hand. Shift work is not unusual. Night work is not unusual. Overtime work can be necessary with little or no notice. Very often the companies want the workers to stay and remain dedicated to the job, as if they had no family at home. Sometimes this is a very unreasonable request. It's not impossible, though. Here are some tips.

1. **Because of the early starts in construction it is difficult to arrange preparing children to go to school.** Very often the parent is gone to work before the children need to leave for school. Working with a partner, family, and friends to provide assistance is key.

2. **Schedules, shift work, and overtime do not take into account daily routines** for school and childcare. Ideally, flexible back up including a partner, family, friends, and paid babysitters can help ease the burden.

3. **On the positive side, since construction work is usually done early in the day, workers often find the opportunity to become involved with their children's after-school activities.** This includes helping to coach sports teams and being able to support other activities that the kids are involved in.

4. **Be aware that companies are not concerned with a worker's home life including relationships and children.** It is best to keep one's private life separate from work.

5. **Sometimes with new babies, women lose a significant amount of time from work caring for their children.** It can be very hard to get re-started in a job. Contacting the union (if you have one) or a supervisor can be helpful in getting re-started.

6. **If you are a parent, remember why you came into construction.** It's a good way to earn a good wage for satisfying work. You are setting a great example for your children.

> **See also:** **Skill/Ability 12:** Determination
> **Success Tip 2:** Stay Tough
> **Success Tip 3:** Build on your Strengths
> **Challenge/Opportunity 15:** Colleagues are aggressive, racist, sexist, or homophobic.

Take Action: One thing that many parents have done is to have a "network" of friends and family who are willing to step in and help with children if work schedules change without notice. Talk to other parents who have to juggle parenting and busy schedules to get some tips—and some support.

CHALLENGE/OPPORTUNITY 8

Obtaining a Mentor

Sometimes we all need help. A mentor is someone who is committed to your growth and professional development. Typically, mentors help you learn what you need to know on the job and how to improve.

1. **Mentors in construction are hard to find.** Usually they come in the form of a supervisor, or more experienced worker who "takes you under their wing" to teach you the trade.

2. **In terms of guidance for life situations,** it is very rare for construction workers to become close enough to share their personal business. Stick to work as much as possible.

3. **A good first step to finding a mentor is to figure out what exactly you would like help with:** are you looking for general professional advice, how to handle sticky situations, how to improve your technical skills, how to advance or get a raise, or something else? Then you can start looking for someone who is especially knowledgeable about that area.

4. **See if your company, union, or other association has a mentoring program, new employee orientation, or early career networking meetings.** Those are excellent places to find mentors.

5. **Ideally, your mentor has your best interests and professional development in mind,** but they are human too and may be interested in learning information from you or steer you in certain directions. If you feel your mentor is not acting in your best interests, you may want to step back and reassess.

6. **Respect your mentor's time.** Be on time and prepared for meetings. Do your homework first (don't ask them to tell you things a simple Internet search could uncover). You may want to bring a written list of what you want to discuss.

7. **Follow up on their advice.** A good mentor will always allow for the mentee to make their own judgments, without prejudice. A good mentor will let you run and will help you out.

8. **There are few things a mentor loves to hear** more than "Your advice was so helpful. It really made a difference."

9. **Be aware that it's rare to find a single mentor who can meet all of your needs.** You may find one person who can help you with general professional advice, and another one who can advise on advancement. Go slowly on questions and ask your mentor how they would like to proceed.

10. **Some mentorships last a lifetime, others for a short time.** If you need to end a mentoring relationship for whatever reason, it's always polite to thank the person for being there for you and let them know how much they have helped you.

11. **Help your mentor out when you can.** Share non-confidential information they might find helpful. Ask them directly how you can help them.

12. **If you find a mentor isn't particularly helpful,** be specific and ask for what you are looking for. Mentors want to be helpful and will likely let you know if they can't help on a specific topic.

13. **Pass it on.** Regardless of your position and seniority, you know something someone doesn't. Be kind and mentor others. It will help you build positive relationships, and it will make the world a little friendlier for all of us.

See also: **Skill/Ability 12:** Determination
Success Tip 3: Build on your Strengths
Success Tip 11: Take Initiative
Success Tip 13: Always Be Learning More

Take Action: People provide mentoring in different ways. Observe any mentoring that may be happening around you and take a broad perspective: If someone says something helpful, consider it mentoring.

CHALLENGE/OPPORTUNITY 9

Interviewing for a Job in the Trades

Some construction companies will conduct an application process similar to other formal employment by advertising, requesting a resume, and conducting an interview. Many other companies, however, do not have a formal process. They post for help wanted and hire those who show up. And unions work differently too: you apply and interview to get into the union and then the referrals to contractors come from the union.

1. **Very often, Millennials that are new to construction are hired through knowing someone else who has vouched for them as someone with a good attitude and a willingness to work.** Having a good reference to start a job is always a good thing. A lot of times it's about who you know to get started. Being able to talk about who you trained and apprenticed with is important. Remember your reputation may very well get to the job before you do.

2. **The ones who actually hire the crew are usually the foreman or superintendents in the field.** They are authorized by the bosses running the company to hire workers as needed.

3. **Construction job interviews are short—usually about a minute.** The interview process is concise, because the foremen running the work will evaluate the newly hired person very quickly based on if the proper tools and equipment have been brought and if the person has the experience to do the job or is

teachable. Through the initial contact, attitude toward work and enthusiasm are quickly sized up. If the foreman thinks he has someone who is willing to do a day's work, then the job interview might turn into employment. They need to get their jobs done, so they look for who will be productive.

4. **Prepare for questions you will likely be asked and practice how you will answer them.** Common interview questions include: (1) What is your experience or training? (2) What are you prepared to do today? (3) When can you start?

5. **Sometimes, a foreman/forewoman may "try" you for a short time.** Within an hour they can see if you know what you are doing or not. If you are not what you presented or not what they need, it might be a very short-term job. (As in: "Don't come back tomorrow!") Also, jobs can end abruptly. You may be hired just to help finish off a project.

6. **Realize that every time you go to work, you gain five things**: (1) you get paid to do work; (2) you gain more valuable experience in the way things are done; (3) you become better practiced at your trade; (4) you open up an opportunity to make contacts with contractors and workers which may benefit you in the future; and (5) you may learn new ways to do things. Even if the job seems to be short-term, it is important to go and take advantage of the opportunity for even a small success. Small successes added together become larger overall successes.

7. **Show up early!** It's a great idea to plan to arrive an hour early and do any final preparation at a coffee shop around the corner if that's possible. This also gives you some cushion in case anything goes wrong on your commute. Do not be late.

8. **Make as many friends as you can.** When you make good relationships, they can mean a lot, especially when looking for work. If you cross someone, or let them down, that relationship will end.

See also: **Skill/Ability 12:** Determination
Success Tip 2: Stay Tough
Success Tip 3: Build on your Strengths

Take Action: Now and for the rest of your career in construction, have a list of contacts in companies or people who can refer you to work. You may start with just a few people you met at training or apprenticeship. Work up to at least 100. When work slows down, you may need to call all of them to get one or two leads.

CHALLENGE/OPPORTUNITY 10

Starting a New Job in the Trades

Great! You have a new job! What can you do to prepare and to get started on the right foot? Keep reading . . .

1. **First of all, be on time!** In construction being on time is always arriving at the jobsite at least twenty to thirty minutes early.

2. **Dress properly.** Look like you are dressed to do the type of work expected.

3. **Be prepared with all the necessary tools you are expected to bring.** If you are brand-new or if you don't know what to bring, at least bring a tape measure, a pencil and a small notebook.

4. **Introduce yourself to your foreman or supervisor using this line:** "Hi, my name is___. What would you like me to do first?" This gives the foreman the idea that you are there to work and be productive. It usually works, and it's a great way to start off on a new job. Once you have been assigned to do something, get busy and go about the task assigned. (Even if you don't particularly like the task.)

5. **First impressions are extremely important.** Be sure the foreman/forewoman knows you are there to be productive. In the first minute of arriving at the job, the foreman is already making up his mind if you are going to be kept or not.

6. **Put your best foot forward as you start the work.** Be busy and work hard to be productive as you start off with a company. They are evaluating you from the very beginning.

7. **Introduce yourself kindly and generously to everyone you meet.** Let people know your name, job title, and your manager or the team you're working on. At this point, you don't know who they are and you never know where they will be in the future.

8. **Be prepared to answer questions about where you were before you started this job.** If you took time off or had a difficult experience before starting this job, practice what you will say so that you can say it breezily. Don't overshare.

> **See also:** **Skill/Ability 12:** Determination
> **Success Tip 1:** Safety first
> **Success Tip 3:** Build on your Strengths
> **Success Tip 13:** Always Be Learning More

Take Action: If you've had a job before, think back to how you managed your first day. What can you do better next time? If you haven't had a job, talk with someone who has and ask them how to best handle the first day at a new job.

Building Relationships with Colleagues

Getting along with others is essential, especially on a construction site. Construction people come in all shapes and sizes with every background imaginable. The one thing they all have in common is they are all trying to earn a living. Remember that you are at the jobsite to earn a living, and that's a lot easier if you get along.

1. **It is a good idea to get along with the others who are long-time workers with a company.** Very often they will be asked by a supervisor about how a newer person is doing on a job. Most will give a fair representation.

2. **Do not get involved in stories about other workers.** Do not gossip. Mind yourself and be accountable for your own actions.

3. **When working with a partner or a group, work as a team.** You will all sink or swim together, and it is frowned upon to blame your partner or others in your work gang for failures or mistakes. The foreman/forewoman may not take note of the blame, but the other workers will.

4. **One of the best things any worker can do is to go to work and focus on the job at hand.** It sounds like a simplification, but it is true.

5. **Do your work to the best of your ability and do the work to the set standard in a timely fashion.**

6. **Approach work with the positive attitude** that "We're all here to get a job done, and it's easier and less stressful if we work together." Maintaining a positive attitude will also serve you well throughout life! This will keep you employed until the end of the project and set the tone to keep you if there is more work available with that company.

7. **Be cautious about socializing with your team or your boss outside of work.** This depends greatly on the work context. In many organizations, an occasional happy hour is fine, but any more than that might be seen as inappropriate.

8. **It's important to find allies at work.** Start by offering support to your colleagues when you notice they need help. This will develop trust over time. Make sure you don't ask for too much and that you help them out when you can.

9. **Manage boundaries at work appropriately.** Friendships can impact your job either because you are spending so much time together or because there is a disagreement. Either way, make sure your focus at work is on work.

10. **Trust is critical in all relationships**, including work relationships. If you mess up, fess up and work on doing better next time.

11. **Help others find their greatness.** If you support and encourage your colleagues to achieve their potential, everyone benefits.

12. **As a Millennial on social media, may want to keep it separated into personal and professional accounts and do not follow anyone you work with on your personal accounts.** If you are on social media at work, don't assume content is private or will be kept private, regardless of the privacy settings you chose.

See also: **Success Tip 6:** Show up with a Positive Attitude
Challenge/Opportunity 14: Socializing Outside Work
Challenge/Opportunity 15: Colleagues Don't Contribute

Take Action: Tomorrow, go up to one person and just say hi (or nod). Work up to actually talking with someone. You can do this.

Socializing outside of Work—especially as a Millennial

We have advised you to keep work at work and keep your private life outside of work. But what do you do when a colleague or boss asks you to get together outside of work?

1. **In many places it is common for workers to go to a bar for a few drinks on the way home from work.** It is common to kick back and relax after a hard day. Sometimes good relationships are established over a few beers. Don't share anything negative about the company, the job, or your coworkers or you may find it getting back to them.

2. **Similarly, be careful who you drink with.** False information and rumors are fueled by alcohol. Things can get out of hand very quickly. Emotions can run high and offense can easily be taken. The rules of the jobsites do not transfer over to the bar, but the trouble started in a bar can often carry over to the job days later.

3. **If someone asks you out for a social drink or for a date and you don't want to go out with them, be explicit and clear in your "no."** If you leave room for doubt, the person may misunderstand and ask again.

4. **You can let your guard down at a bar—somewhat**. Even after hours, it's not a good idea to engage in sexual or raunchy conversations, as these have a way of getting back to your employer.

5. **Be cautious about socializing with your boss outside of work**. This depends greatly on the work context. In many organizations, an occasional all-office happy hour is fine, but any more than that might be seen as inappropriate.

6. **Don't party too hard.** Many employers fire workers who show up inebriated or high. Others drug test employees. It's not worth your job!

7. **Manage boundaries at work—and when socializing—appropriately. Friendships can start to impact your job either because you are spending so much time together or because there is a disagreement. Either way, make sure your focus at work is on work.**

See also: Success Tip 8: Be Honest and Earn Trust
Challenge/Opportunity 8: Obtaining a Mentor
Challenge/Opportunity 11: Building Relationships with Colleagues

Take Action: Identify people in your training school or jobsite who are respected and responsible. Consider following their lead for socializing outside work.

Colleagues don't Contribute

From grade school projects to day-to-day work, sometimes our colleagues just don't pull their own weight. No need to get angry—just try some strategies to help things move forward.

1. **Start by assuming that your colleague has something else getting in the way of completing their work** as opposed to they are just refusing to work. Check with them and see if there is something else going on.

2. **If you are working with a partner or as part of a group, realize that it is a bad idea to make others look bad, or blame them for some flaw.** When questioned about something that is wrong, own up to it by saying: "We made a mistake, and we will correct it." If you blame others, they will never forget, and they will always hold that against you.

3. **Ask the colleague if they are having trouble completing their part of the work and why.** There might be a way to help them, suggest resources, or otherwise address the problem while letting everyone save face.

4. **Set clear boundaries and be sure not to take on the other person's work responsibilities.** Being professional and thoughtful doesn't mean you do their work. Find that balance.

5. **If you must involve your boss, such as if your efforts aren't**

working or if your colleague is creating a safety problem, consider asking a colleague or mentor not involved with your work site how to handle the situation, without naming names. Go to your boss only when necessary (such as a safety issue).

6. **If this is a pattern with this particular colleague not doing their work**, try to get work on a project not working with this colleague or at another jobsite.

See also: Success Tip 6: Show up with a Positive Attitude
Success Tip 9: Focus on Getting Stuff Done
Challenge/Opportunity 13: Building relationships with colleagues

---

Take Action: Who can you talk to about a problem like this? Find a mentor, friend, or other person you respect you can discuss this with.

CHALLENGE/OPPORTUNITY 14

Asking for a Raise or Promotion

We all want to get paid more, but raises often seem elusive. Asking your boss for a raise is one of the most stressful conversations to have. But you can do it and be successful!

1. **Proactively communicate your accomplishments to your boss over time** so that they have a sense of the good work you have been doing.

2. **Be sure you deserve a raise and that you can document your accomplishments.** Gather your evidence, keeping in mind that often bosses prefer quantitative accomplishments (1% error rate, the lowest in the group) compared to qualitative accomplishments (minimal complaints). Consider writing out your arguments first before you go to the boss.

3. **Identify whether there is a good time to ask for a raise.** Sometimes it's good to ask for a raise as bosses are preparing next year's budget or when a project is starting or ending. Check with a trusted colleague or mentor to consider timing.

4. **Identify the circumstances under which raises are given.** If your company provides raises only after stellar performance reviews, an off-season raise would be challenging for your boss to approve.

5. **You may want to practice this conversation with a friendly partner beforehand.** Anticipate what your boss might say

and practice how you would respond to various comments. Encourage your partner to keep it as realistic as possible; let them know what you are worried or afraid your boss might say, and practice how you respond.

6. **Make sure your reasons for deserving a promotion are professional, not personal.** Do not mention that you are buying a new house or that your mother thinks you should be making more. These are not relevant. Focus on the value you provide and what you bring to the organization. Remember, your boss may have to justify the raise to their bosses; give them enough information to do so successfully.

7. **Clarify to your boss how you are committed to the organization and will help it grow in the future.**

8. **If you have data, such as from your organization or from the field,** that shows you are being underpaid relative to your skills and experience, consider discussing a raise as an opportunity to achieve equity with salaries.

9. **If your boss says no to a raise,** ask for how to improve your performance, or what your boss would like to see before they would approve a raise. You can also ask for an alternative, such as additional paid time off or funding to attend a conference.

10. **If your boss says maybe or they will think about it,** make sure you clarify next steps. You could say something like, "Thanks for hearing me out. Would it be okay for me to check back in with you when we meet again in two weeks?"

11. **Be aware that some industries only give raises when you have an offer from another organization and are essentially threatening to leave**. If you go this route, be sure you are willing to leave if you don't get the raise. Work with a trusted colleague, mentor, or coach on timing and framing of these conversations.

See also: **Skill/Ability 7:** Logical Thinking and Planning
Skill/Ability 9: Ability to Overcome Fears
Skill/Ability 11: Uncertainty in Employment

Take Action: Be clear on your values and know what you're worth. Talk with a friend, mentor, or other trusted person to clarify your approach.

CHALLENGE/OPPORTUNITY 15

Colleagues are Aggressive, Racist, Sexist, or Homophobic

We hope you never have to deal with aggressive, racist, sexist, homophobic, or hostile colleagues. If you do, we want you to have tools to deal with the situation in the moment and to take it to the appropriate place at your organization, so it stops.

1. **It shouldn't feel necessary, but state clearly what you are requesting.** For example, "Please don't use that word around me. That's not right."

2. **If the person claims they are kidding around,** you can say, "I don't like jokes about [x]. Please stop."

3. **If it's possible to confront someone privately, you are more likely to have a productive conversation with the person.** If there are others around who are being affected by the comment or action, you may want to consider standing up in front of others intentionally, especially if you know they will have your back.

4. **Consider the difference between a request and a demand.** You can start with a request ("Please don't use that word") and then escalate to a demand if needed. For example, "Jokes about women are not funny, and I won't listen anymore." Then walk away.

5. **Document what happened including the date, location,**

people present, and what happened, even if in your own file. Have "receipts" you can go back to.

6. **Get support from others, including friends or family.** Be cautious discussing the situation too broadly within the organization, especially if it could possibly lead to legal action for discrimination.

7. **If coworkers are being loud and obnoxious, you can attempt to reign in the conversation.** For example, "Let's get back to the work. We were about to ..."

8. **It's important to speak up when you hear inappropriate comments.** If you don't speak up, it often gives the impression that you are condoning them even if you are not.

9. **Consider contacting a mentor, colleague (outside the organization), union representative, or boss** if discriminatory language or actions do not stop. If you need to find a lawyer, contact a local university law school for an inexpensive or low-cost referral or Google "best employment lawyer" with the name of your town and call to ask for a free consultation

---

See also: Success Tip 2: Stay Tough
Success Tip 8: Be Honest and Earn Trust
Success Tip 14: Show Solidarity

Take Action: If you're experiencing aggression, racism, sexism, homophobia, or hostility at work, take at least one step above. Do not just let it be, because it will not get better on its own.

CHALLENGE/OPPORTUNITY 16

What to do When you Get Promoted

You got the promotion! Congratulations! Now you're the foreman, forewoman, supervisor, or boss of . . . people who were previously your coworkers or buddies. This could be awkward. And how do you supervise anyway?

1. **Don't let it go to your head.** Your employees want to be led, not managed to the extreme. You know what to do for a job, so share with the others what they need to do, and let them do it. Be confident in what they're doing and what you're doing.

2. **You may want to consider discussing with your mentor about starting a new job** as a boss, including goal setting, network building, and achieving early wins.

3. **Be prepared to answer questions about where you trained and your experience before you started this job.** If you took time off or had a difficult experience before starting this job, practice what you will say so that you can say it breezily until you get to know people better.

4. **Remember the individuals who interviewed you for the promotion** and when you see them at work, say something positive.

5. **Consider how you'll introduce yourself to your new workers,** which may depend on the kind of promotion you received. You

can introduce yourself casually, have a meeting for a more formal introduction, or keep on moving without saying anything.

6. **People may be curious about you,** including why you got promoted, how you want to approach work, and what your strengths and weaknesses are. Know that you're being observed, maintain boundaries with what you want to share, and consider how you want to respond to inappropriate questions.

7. **Allow time for reflection.** It's easy to hit the ground running, but you want to make sure you're headed in the right direction! At least once a week, take some time to review what you've learned and ensure you're on the right track.

8. **Reward yourself with something pleasant after you finish your first day,** whether dinner with your partner or friends, a special dessert, or something else. You did it!

9. **Check out** *Millennials' Guide to Management & Leadership* by Jennifer P. Wisdom. It's another Millennials' Guide that might have just what you need.

See also: **Success Tip 2:** Stay Tough
Success Tip 8: Be Honest and Earn Trust
Success Tip 14: Show Solidarity

Take Action: Consider great bosses and teachers you have worked with in the past. What made them great? Identify your own path to becoming a great boss.

CHALLENGE/OPPORTUNITY 17

You Think it's Time to Move on

You might be ready to start looking for another job. But how do you do that when you're busy at work? How do you know if you should stay or go? How do you determine when you should tell your boss, especially if you need a reference? The good news is that many construction companies expect employee turnover to some extent and will not be shocked that an employee is leaving. The bad news is that our bosses are all still just people, who sometimes feel hurt or angry that you're leaving. Here are some ways to navigate this process.

1. **In construction as in any field, there may come a time when you need to leave a company.** You may be stuck in a rut of doing the same type of work over and over. It is also possible that you want to try to move up to supervisory level, and the current situation does not allow for it. There are many reasons to leave.

2. **Be aware, however, that in construction some companies may hold your departure against you!** You may never work for that company again. It is just how they are, and it's not personal. Keep that in mind if you plan to change form one company to another. Always try to leave on good terms as best you can.

3. **Be very clear on the reasons you are considering leaving.** Most Millennials leave organizations because they're frustrated with the work, they aren't receiving feedback, they have a terrible

boss, or they no longer have passion for the work. Why do you want to leave?

4. **Consider whether it makes sense to leave the company or to leave for a different part of the company or a different jobsite.** Generally, leaving an institution makes sense if many or key relationships in the organization are strained or unsupportive; if you're generally dissatisfied with the institution itself or its leaders' major decisions; or if you want a new challenge or new area to live.

5. **Interviewing while employed can be challenging.** If your boss finds out you're interviewing, sometimes they will fire you on the spot. Usually it's most appropriate to take vacation time/paid time off for interviews.

6. **Before you decide to take action on leaving, reassess your goals and priorities.** Is there a way your goals can be met in this position? Get very clear on what you want and what the best way is to get it.

7. **A good time to tell your boss you're looking for another job is either when you start to interview** (if it's a small field and you think word will get back) or when you are accepting a position (if your boss likely doesn't know and you think they could try to derail the offer), depending on your relationship with your boss.

8. **Before you start to leave, identify a story you can share with prospective employers about why you're leaving.** This story should be positive and authentic. You will likely be asked in an interview about why you want to leave your current employer; never say anything bad about them, as the interviewer will assume you will also badmouth your new employer too. Even if you are leaving because your boss is a jerk, ideally you can frame your story about opportunity for growth or contribution at the new organization.

9. **Start identifying projects you may need to wrap up at your**

current organization and any logistical concerns, such as when your retirement contributions are vested or when insurance may end.

10. **Don't burn bridges as you leave a job.** You never know how you may work with this organization or boss or colleagues again. I suggest you start by saying something positive about the job, explain why you're leaving, then help the boss make the transition as easy as possible (such as by being flexible on your last day or describing a transition plan).

11. **After you've left, keep in touch with bosses and colleagues with whom you are on good terms.** Keeping these relationships can be helpful in the future when you—or they—may be looking for other work.

> **See also:** **Success Tip 3**: Build on your Strengths
> **Challenge/Opportunity 9:** Interviewing for a Job in the Trades
> **Challenge/Opportunity 14:** Asking for a Raise or Promotion

Take Action: Ask around at your trade school or apprenticeship (not at your jobsite) how transitions between jobs usually happen in your field. Ask how people left jobs on good terms, and for any advice they have.

Starting your own Business vs. Working for a Company

Many Millennials have a desire to start their own company and to be their own boss. It is part of the Great American Dream! Many construction companies are started by tradespeople and become very successful. There is a lot of money spent on construction, and a lot of money to be made! Should you start your own business? Read on . . .

1. **Being a skilled tradesworker does not guarantee success in business.** Running a business has many variables. The actual work of the trade is only one part of it.

2. **A huge part of running a company is finding work to do that is profitable for the owner.** Pricing construction projects and estimating is very different from doing construction work itself. A lot of effort in running a company is focused on building relationships to obtain contracts that are profitable.

3. **The business side of any business, meaning accounting, budgeting, cost analysis, finances, human resources, and so on are all necessary.** Often, tradesworkers starting a business are very unaware of the business side and thus are unprepared when dealing with the business part of running a company. Millennials starting a construction company, or any other company should take classes in business and management. A basic understanding

of how to write a business plan, business finance, accounting and management are all essential.

4. **For most construction jobs, at the end of the day, you can go home with a clear conscience and forget about the job.** Once you become the "Boss" or the owner, the days become much longer. You are no longer paid by the hour, but rather by the success of the projects completed. Yes, the rewards can be great, but running a company can be exhausting and often the amount of time spent is far greater than a 40-hour work week.

5. **Most Millennials (and everyone else!) want to start a business to make money.** It's important to have an accurate sense of the funds and time needed to get started before the money starts coming in. This can include you spending hours to prepare bids for work, hire workers, pay workers weekly before you receive payment, and spend late nights balancing budgets. As a boss, you have stress and pressure to complete jobs and take care of your workers. Think about the amount of capital funds, your skills, and the kind of lifestyle you want before you get started.

6. **The world of work can be precarious and difficult.** Not everyone in construction shares the same values. Not all contracts are paid, even though they were completed on time and according to specifications. Many people cheat. Some contractors do not pay their workers or their taxes. Sometimes workers are unknowingly working without insurance. Some contractors are thieves and criminals. For someone starting a business, it is hard to figure out who is who, and sometimes by the time bad people reveal their true intentions, it is too late for those involved in the projects. Ensure you have good mentors and check references before starting projects.

7. **Small construction companies often work as subcontractors.** That means there is a general contractor or construction manager in charge of the project, and your company, specializing in plumbing or electrical work for example, works for the

contractors. As above, not all companies pay you fairly, pay their own workers fairly, or have the same high standards for safety as you do. As a subcontractor, you will need to watch out for your interests and the interests of your workers.

8. **Before you start your own business, learn as much as you can about the business itself.** Observe construction managers, foremen/forewomen, and the business side of construction. Consider what skills you have (such as relationship building) and what work you may need to hire someone to do (accounting).

9. **In addition, before you start your own business, find a mentor or two.** You will find that the relationships side of business is very important, and a mentor can assist you in opening doors, developing relationships, and maintaining relationships. Local chambers of commerce can be good sources for finding colleagues and mentors.

10. **Time is irreplaceable.** Once it is gone, it cannot be replaced. Think of how you would like to spend your life.

See also: Success Tip 2: Stay Tough
Success Tip 3: Build on your Strengths
Skill/Ability 7: Logical Thinking and Planning
Skill/Ability 8: Organizational Skills
Skill/Ability 12: Determination

Take Action: If you're thinking about starting your own business, make a list of your strengths. Identify strengths that are specific to the construction work itself, such as mechanical abilities, and which apply to the business world generally, including developing relationships, managing contracts, and understanding regulations. Seek guidance to address any areas you would like to build.

Not Sure Where to Go for Help

If you're stumped with what you think is a completely unique problem, help is on the way! There are lots of ways to identify solutions for your unique set of circumstances.

1. **Google the problem** to see if anyone has addressed it online and identify who they went to for help.

2. **Find online forums related to your problem,** such as Reddit or Quora.

3. **Don't blame yourself if you're not sure how to address a problem;** sometimes there are larger political forces at work of which you are only seeing a small part. Still try to address your part in a way that is consistent with your values.

4. **Identify a colleague who seems to always know where to go with problems and ask for advice.**

5. **Talk to someone you trust outside of work** for an external opinion on the problem. Former trade school instructors or friendly former colleagues could be a good start.

6. **Consider waiting if waiting won't cause irreparable damage;** sometimes problems work themselves out. If you do this, make a note of how long you should wait before you revisit the issue.

7. **Clarify the problem in writing, identify some possible solutions, and then try one.**

8. **Gather some friends and ask them to help you brainstorm solutions**. This also will remind you of what's important, and how much support you have. It will help put the problems in perspective.

9. **You can always email us** at karl@karldhughes.com or jennifer@leadwithwisdom.com. We're happy to help you out!

FOR FURTHER READING

Trade-specific Resources

Boilermaker

◆ National Center for Construction Education and Research (NCCER): *www.nccer.org*

◆ Bureau of Labor Statistics/Occupational Outlook Handbook information: *www.bls.gov/ooh/construction-and-extraction/boilermakers.htm#tab-4*

◆ International Brotherhood of Boilermakers, Iron Ship Builders, Blacksmiths, Forgers and Helpers: *www.boilermakers.org*

◆ American Federation of Labor and Congress of Industrial Organizations and Central Labor Councils (AFL-CIO): *www.aflcio.org*

◆ American Boiler Institute: *www.americanboilerinstitute.com*

Bricklayer/Mason

◆ Bureau of Labor Statistics/Occupational Outlook Handbook information: *www.bls.gov/ooh/construction-and-extraction/brickmasons-blockmasons-and-stonemasons.htm*

- International Union of Bricklayers and Allied Craftworkers (BAC): *bacweb.org/*

- Mason Contractors Association of America (MCAA): *www.masoncontractors.org/history/*

- The Home Builders Institute (HBI): *www.hbi.org*

- The International Masonry Institute: (IMI) *www.imiweb.org*

Cabinetmaker/Woodworker

- Bureau of Labor Statistics/Occupational Outlook Handbook information: *www.bls.gov/ooh/production/woodworkers.htm#tab-1*

- United Brotherhood of Carpenters and Joiners of America: www.carpenters.org

- The Woodwork Career Alliance of North America (WCA): *www.woodworkcareer.org*

Carpenter

- Bureau of Labor Statistics/Occupational Outlook Handbook information: *www.bls.gov/OOH/construction-and-extraction/carpenters.htm*

- United Brotherhood of Carpenters and Joiners of America: *www.carpenters.org*

- Building Works: *www.nyccarpenterstrainingcenter.org/buildingworks*

- Occupational Safety and Health Administration (OSHA): *www.osha.com*

- Pre-Apprenticeship Certificate Training (PACT): *www.hbi. org/Programs/Training-Programs/PACT-Programs*

- Home Builders Institute (HBI): *www.hbi.org*

- National Association of the Remodeling Industry (NARI): *www.nari.org*

Construction and Building Inspector

- Bureau of Labor Statistics/Occupational Outlook Handbook information: *https://www.bls.gov/ooh/construction-and-extraction/construction-and-building-inspectors.htm*

- International Code Council (ICC): *www.iccsafe.org*

- International Association of Plumbing and Mechanical Officials (IAPMO): *www.iapmo.org*

- International Association of Electrical Inspectors (IAEI): *www.iaei.org*

- National Fire Protection Association (NFPA): *www.nfpa.org*

- Occupational Safety and Health Administration (OSHA): *www.osha.com*

- American Federation of Labor and Congress of Industrial Organizations and Central Labor Councils (AFL-CIO): *www.aflcio.org*

Diesel Service Technician

- Bureau of Labor Statistics/Occupational Outlook Handbook information: *https://www.bls.gov/ooh/installation-maintenance-and-repair/heavy-vehicle-and-mobile-equipment-service-technicians.htm*

- National Institute for Automotive Service Excellence (ASE): *www.ase.com*

- International Association of Machinists (IAM) and Aerospace Workers. *www.goiam.org*

Drafter

- Bureau of Labor Statistics/Occupational Outlook Handbook information: *www.bls.gov/ooh/architecture-and-engineering/drafters.htm#tab-1*

- American Design Drafting Association (ADDA): *www.adda.org*

Electrician

- Bureau of Labor Statistics/Occupational Outlook Handbook information: www.bls.gov/ooh/construction-and-extraction/electricians.htm#tab-1

- Pre-Apprenticeship Certificate Training (PACT): *www.hbi.org/Programs/Training-Programs/PACT-Programs*

- National Electrical Contractors Association (NECA): *www.necanet.org*

- International Brotherhood of Electrical Workers (IBEW): *www.ibew.org*

Flooring Installer/Tile & Marble Setter

- Bureau of Labor Statistics/Occupational Outlook Handbook information: www.bls.gov/ooh/construction-and-extraction/tile-and-marble-setters.htm#tab-1

- Ceramic Tile Education Foundation (CTEF): *www. ceramictilefoundation.org*

- International Masonry Institute (IMI): *www.imiweb.org*

- International Union of Bricklayers and Allied Craftworkers (IUBAC): *www.bacweb.org*

- National Tile Contractors Association (NTCA): *www.tile-assn.com*

- Tile Contractors' Association of America (TCAA): *www.tcaainc.org*

- Tile Council of North America (TCNA): *www.tcnatile.com*

- Advanced Certifications for Tile Installers (ACT) program: *www.ceramictilefoundation.org/ advanced-certifications-for-tile-installers*

- National Wood Flooring Association (NWFA): *www.nwfa.org*

- International Certified Floorcovering Installers Association (CFI): *cfiinstallers.org*

- International Standards & Training Alliance (INSTALL): *www.installfloors.org*

- International Union of Painters and Allied Trades (IUPAT): *www.iupat.org*

Glazier

- Bureau of Labor Statistics/Occupational Outlook Handbook information: *www.bls.gov/ooh/construction-and-extraction/ glaziers.htm#tab-1*

- International Union of Painters and Allied Trades (IUPAT): *www.iupat.org/join-us/glazing*

Heating, Ventilation, Air-conditioning, and Refrigeration (HVAC-R) Technician

◆ Bureau of Labor Statistics/Occupational Outlook Handbook information: *www.bls.gov/ooh/installation-maintenance-and-repair/heating-air-conditioning-and-refrigeration-mechanics-and-installers.htm#tab-1*

◆ United Association of Journeymen and Apprentices of the Plumbing and Pipe Fitting Industry of the United States, Canada (UA): *www.ua.org*

◆ The International Association of Sheet Metal, Air, Rail and Transportation Workers (SMART): *www.smart-union.org*

◆ U.S. Environmental Protection Agency (EPA): *www.epa.gov*

Heavy Equipment Operator

◆ Bureau of Labor Statistics/Occupational Outlook Handbook information: *www.bls.gov/OOH/construction-and-extraction/construction-equipment-operators.htm#tab-1*

◆ International Union of Operating Engineers: *www.iuoe.org/*

Ironworker

◆ Bureau of Labor Statistics/Occupational Outlook Handbook information: *www.bls.gov/ooh/construction-and-extraction/structural-iron-and-steel-workers.htm#tab-1*

◆ International Association of Bridge, Structural, Ornamental, and Reinforcing Iron Workers: www.ironworkers.org

◆ International Brotherhood of Boilermakers, Iron Ship Builders, Blacksmiths, Forgers and Helpers: *www.boilermakers.org*

- American Welding Society: *www.aws.org*
- National Commission for the Certification of Crane Operators: *www.nccco.org*
- National Center for Construction Education and Research: *www.nccer.org*

Laborer

- Bureau of Labor Statistics/Occupational Outlook Handbook information: *www.bls.gov/OOH/construction-and-extraction/construction-laborers-and-helpers.htm#tab-1*
- Laborers' International Union of North America (LIUNA): *www.liuna.org*

Machinist/Tool and Die Maker

- Bureau of Labor Statistics/Occupational Outlook Handbook information: *https://www.bls.gov/oes/current/oes514111.htm*
- International Association of Machinists and Aerospace Workers: www.goiam.org
- Skills Certification System: *www.themanufacturinginstitute.org/workers/skills-certifications*

Millwright

- Bureau of Labor Statistics/Occupational Outlook Handbook information: www.bls.gov/ooh/installation-maintenance-and-repair/industrial-machinery-mechanics-and-maintenance-workers-and-millwrights.htm#tab-1

◆ United Steelworkers (USW): *www.usw.org*

Painter

◆ Bureau of Labor Statistics/Occupational Outlook Handbook
information: *www.bls.gov/ooh/production/painting-and-coating-
workers.htm#tab-1*

◆ International Union of Painters and Associated Trades
(IUPAT): *www.iupat.org*

Plasterer/Stucco Mason

◆ Bureau of Labor Statistics information: *www.bls.gov/oes/2017/
may/oes472161.htm*

◆ Operative Plasterers' and Cement Masons' International
Association (OPCMIA): *www.opcmia.org*

Plumber, Pipefitter, and Steamfitter

◆ Bureau of Labor Statistics/Occupational Outlook Handbook
information: *www.bls.gov/OOH/construction-and-extraction/
plumbers-pipefitters-and-steamfitters.htm#tab-1*

◆ United Association of Journeymen and Apprentices of the
Plumbing and Pipe Fitting Industry of the United States,
Canada and Australia (UA): *www.ua.org*

Roofer

◆ Bureau of Labor Statistics/Occupational Outlook Handbook

information: *www.bls.gov/ooh/construction-and-extraction/roofers.htm#tab-1*

◆ United Union of Roofers, Waterproofers, & Allied Workers: *www.unionroofers.com*

Sheet Metal Worker

◆ Bureau of Labor Statistics/Occupational Outlook Handbook information: *www.bls.gov/ooh/construction-and-extraction/roofers.htm#tab-1*

◆ The International Association of Sheet Metal, Air, Rail and Transportation Workers: *www.smart-union.org*

◆ American Welding Society (AWS): *www.aws.org*

◆ International Certification Board (ICB): *www.icbcertified.org/site/home/index.php*

◆ Fabricators & Manufacturers Association, International (FMA): *www.fmamfg.org*

Surveyor

◆ Bureau of Labor Statistics/Occupational Outlook Handbook information: *www.bls.gov/ooh/architecture-and-engineering/surveyors.htm#tab-1*

◆ International Union of Operating Engineers (IUOE): *www.iuoe.org*

◆ Accreditation Board for Engineering and Technology (ABET): *www.abet.org*

◆ National Council of Examiners for Engineering and Surveying (NCEES): *www.ncees.org*

◆ Fundamentals of Surveying (FS) exam: *www.ncees.org/surveying/fs*

◆ Principles and Practice of Surveying (PS) exam: *www.ncees.org/surveying/ps*

Telecommunications Technician

◆ Bureau of Labor Statistics/Occupational Outlook Handbook information: *www.bls.gov/ooh/installation-maintenance-and-repair/telecommunications-equipment-installers-and-repairers-except-line-installers.htm#tab-1*

◆ Communications Workers of America: *www.cwa-union.org*

Tractor Trailer Truck Drivers

◆ Bureau of Labor Statistics/Occupational Outlook Handbook information: *https://www.bls.gov/ooh/transportation-and-material-moving/heavy-and-tractor-trailer-truck-drivers.htm*

◆ International Association of Sheet Metal, Air, Rail and Transportation Workers (SMART): *www.smart-union.org*

Construction Manager

◆ Bureau of Labor Statistics/Occupational Outlook Handbook: *https://www.bls.gov/oes/2018/may/oes119021.htm*

◆ Construction Management Association of America: *www.cmaanet.org*

◆ American Institute of Constructors: *www.professionalconstructor.org*

- Certified Construction Manager (CCM): *www.cmaanet.org/certification/ccm*

- Associate Constructor (AC): *www.professionalconstructor.org/page/AC_Certification*

- Certified Professional Constructor (CPC): *www. professionalconstructor.org/page/CPC_Certification*

Women in the Trades

- National Association of Women in Construction (NAWIC): *www.nawic.org*

- National Association of Black Women in Construction (NABWIC): *www.nabwic.org*

- Nontraditional Employment for Women (NEW): *www.new-nyc.org*

- Professional Women in Construction (PWC): *www.pwcusa.org*

- Chicago Women in the Trades: *www.cwit.org*

- Apprenticeship & Nontraditional Employment for Women (ANEW): *www.anewaop.org*

- Freemasonry for Women (HFAF): www.hfaf.org

- National Institute for Women in Trades, Technology & Science (iWiTTS): *www.iwitts.org*

- StartZone: *startzone.highline.edu*

- Sisters in the Building Trades: www.sistersinthebuildingtrades.rocks

- The National Taskforce on Tradeswomen's Issues: *www.tradeswomentaskforce.org*

- Missouri Women in Trades: *www.mowit.org*

- Oregon Tradeswomen: *www.tradeswomen.net*

- Policy Group on Tradeswomen's Issues: *www.policygroupontradeswomen.org*

- Tradeswomen Inc: *www.tradeswomen.org*

- Vermont Works for Women: *www.vtworksforwomen.org*

- Washington Women in Trades: *www.wawomenintrades.com*

- Women Contractors Association: www.womencontractors.org

- Women in Non Traditional Employment Roles (WINTER): *www.winterwomen.org*

- Coalition of Labor Union Women (CLUW): *www.cluw.org*

- Minority and Female Skill Trades Association: *www.minorityfemaletradeassociation.com*

- Women in HVACR: *www.womeninhvacr.org*

- Central Ohio Women in the Trades: *www.womeninthetrade.com*

- Women in Construction pre-apprenticeship training *www.moorecommunityhouse.org/winc*

Underrepresented Racial Minorities in the Trades

- U.S. Department of Labor: *www.dol.gov*

- National Hispanic Construction Association: *www.builtbylatinos.org*

- National Association of Black Women in Construction (NABWIC): *www.nabwic.org*

- National Black Contractors Association: *www.nationalbca.org*

- National Association of Minority Contractors (NAMC): *www.namcnational.org*

- StartZone: *startzone.highline.edu*

- Minority and Female Skill Trades Association: *www.minorityfemaletradeassociation.com*

LGBTQ in the Trades

- Pride at Work: *www.prideatwork.org*

- The National LGBT Chamber of Commerce: *www.nglcc.org*

- Build Out Alliance: *www.www.buildoutalliance.org*

Veterans in the Trades

- Helmets to Hardhats: www.helmetstohardhats.org

- Military Skilled Trades: *www.militaryskilledtrades.com*

- Troops to Trades: *www.ieci.org/troops-trades*

Person with a Criminal Conviction in the Trades

- General Education Diploma: *www.ged.com*

- U.S. Department of Labor: *www.dol.gov*

Not sure where to go for help

- Reddit: *www.reddit.com*

- Quora: *www.quora.com*

Organizations to learn more about the trades

Occupational Outlook Handbook

◆ Not sure what career to go into? The Occupational Outlook Handbook has many ways to find career information. *www.bls. gov/ooh/*

Organizations to facilitate entry in the trades

The ACE Mentor Program serves high school youth who are exploring careers in Architecture, Construction, or Engineering. The mentors are professionals from leading design and construction firms who volunteer their time and energy. The program is designed to engage, inform, and challenge youth. *www.acementor.org*

The American College of the Building Arts (ACBA) is dedicated to educating the next generation of building artisans and to preserving the building arts in a manner never before seen in America. Under the direction of our experienced faculty, students have the opportunity to receive a quality liberal arts education while they learn the skills needed to excel in their chosen field. *www.buildingartscollege.us/*

Building Works is a NYC Department of Labor approved pre-apprenticeship program for individuals from underrepresented communities. The program is free and focuses on career guidance, introduction to apprenticeship, worker health & safety, and environmental worker training. Participants may also receive a variety of formal certifications specific to the industry, such as: OSHA construction safety, scaffold awareness, and hazardous waste worker training. *www.nyccarpenterstrainingcenter.org/buildingworks*

American Council for Construction Education (ACCE)—Since 1974, the American Council for Construction Education (ACCE) has been a leading global advocate of quality construction education that promotes, supports, and accredits quality construction education programs. *www.acce-hq.org*

American Road & Transportation Builders Association (ARTBA) has 5 different scholarships available: The Highway Worker Memorial Scholarship Program; the annual Young Executive Development Program; the annual Globe Awards Program; the annual Roadway Work Zone Safety Awareness Awards Program; and the annual ARTBA Student Paper Competition. *www.artba.org/foundation/*

Helmets to Hardhats is a fast way for military, reservists, and those in the National Guard to transition from active duty to a career in the construction industry. *www.helmetstohardhats.org/*

National Association for Women in Construction (NAWIC) helps women take advantage of the opportunities in construction. Whether you want to embark on a new career, establish a networking base, be a mentor/mentee, make a difference in your community, continue your education, or invest in great friendships, NAWIC offers a variety of opportunities—large and small. *www.nawic.org*

National Center for Construction Education & Research (NCCER)—NCCER offers curricula in over 70 different craft areas and more than 80 different assessments. When you successfully complete training, assessments and/or performance verifications through an NCCER Accredited Training Sponsor or Assessment Center, NCCER's Registry System records your completions and issues the appropriate credentials. It is these portable, industry-recognized credentials that many industry leaders look for when making employment decisions. *www.nccer.org/training-and-certifications*

American Federation of Labor and Congress of Industrial Organizations and **Central Labor Councils** is the largest federation of unions in the United States. It is made up of fifty-five national and international unions, together representing more than 12 million active and retired workers. *www.aflcio.org*

ABOUT THE AUTHORS

Karl Hughes is a teacher, mentor, and speaker who uses his background and knowledge to motivate others to personal, professional, and financial success. Karl's 45 years of construction industry experience as a master carpenter, business owner, union member and trade instructor give him a unique perspective on goal setting, action plans, and using your own resources to succeed. Karl has a keen ability to communicate his passions to others.

A sixth-generation carpenter, Karl believes that the construction trades are an overlooked means to a successful life, and he encourages others to explore the opportunities in a trade career. He speaks to his fellow tradesmen and women directly in *Hammering Out a Living*, his Amazon best-selling actionable success journal based on goal setting.

Karl spent more than 25 years volunteering with youth programs. For the past 10 years as a carpentry instructor he has been guiding newcomers to the construction field. This has given him a wealth of experience to share with young people starting out in their careers. He considers it his mission to share the benefits of this alternate route to success in life. He can be reached at www.karldhughes.com.

Jennifer Wisdom, PhD MPH, is a former academician who is now an author, consultant, and speaker and principal of Wisdom Consulting. As a consultant, she helps curious, motivated, and mission-driven professionals to achieve their highest potential by identifying goals and then providing them with the roadmap and guidance to get there.

Jennifer created the best-selling Millennials' Guides series. The first book, *Millennials' Guide to Work*, helps Millennials and others achieve success and respect at work. The second book, *Millennials' Guide to Management & Leadership*, helps Millennials be successful and impactful managers and leaders. She was inspired through her multiple home remodeling adventures to partner with Karl to write about the construction trades.

Dr. Wisdom is a licensed clinical psychologist. She has worked with complex health care, government, and educational environments for 25 years, including serving in the U.S. military, working with non-profit service delivery programs, and as faculty in higher education. She is an intrepid adventurer based in New York City and Portland, Oregon. She can be reached at www.leadwithwisdom.com.

CPSIA information can be obtained
at www.ICGtesting.com
Printed in the USA
LVHW021730080621
689564LV00018B/760